SENTIENT

SENTIENT

✳

HOW ANIMALS ILLUMINATE
THE WONDER OF OUR HUMAN SENSES

JACKIE HIGGINS

ATRIA BOOKS

NEW YORK   LONDON   TORONTO   SYDNEY   NEW DELHI

ATRIA
BOOKS

An Imprint of Simon & Schuster, Inc.
1230 Avenue of the Americas
New York, NY 10020

First Atria Books hardcover edition February 2022

ATRIA BOOKS and colophon are trademarks of Simon & Schuster, Inc.

For information about special discounts for bulk purchases, please contact Simon & Schuster Special Sales at 1-866-506-1949 or business@simonandschuster.com.

The Simon & Schuster Speakers Bureau can bring authors to your live event. For more information or to book an event, contact the Simon & Schuster Speakers Bureau at 1-866-248-3049 or visit our website at www.simonspeakers.com.

Interior design by Kyoko Watanabe

Manufactured in the United States of America

1 3 5 7 9 10 8 6 4 2

Library of Congress Cataloging-in-Publication Data is available on file.

ISBN 978-1-9821-5655-8
ISBN 978-1-9821-5657-2 (ebook)

To my mother,
for sharing her sense of wonder with me

CONTENTS

INTRODUCTION

W E HUMANS ARE OFTEN DESCRIBED AS SENTIENT BEINGS, BUT what does this mean? The word, from the Latin *sentire*, to feel, is so mercurial that the philosopher Daniel Dennett has, perhaps playfully, suggested, "Since there is no established meaning . . . we are free to adopt one of our own choosing." Some use *sentience* interchangeably with the word *consciousness*, a phenomenon that in itself is so elusive as to reduce the most stalwart scientific mind to incantations of magic. Marveling at how brain tissue creates consciousness, how material makes immaterial, Charles Darwin's staunch defender T. H. Huxley once pronounced it "as unaccountable as the appearance of Djin [*sic*] when Aladdin rubbed his lamp." More recently, while probing the soft jelly of a patient's brain, the neurosurgeon Henry Marsh agreed that the idea his fine sucker was passing through thoughts and feelings was "simply too strange to understand." To some scientists, therefore, sentience becomes a hard—if not the hardest—problem in the study of the natural world. However, there is a simpler definition: sentience also describes our ability to sense the world around us. Such sensitivity leads to our experiences of seeing the creamy white page of this book, feeling its weight in our hand, perceiving the murmur of a page turning, but sentience is then the foundation on which the mirage of consciousness shimmers. Scientists and philosophers debate whether animals experience consciousness, but most readily ascribe to them the pared-down version of sentience. This book reflects on how each of the sentient beings with whom we share the planet offers a different perspective on how we sense, even make sense, of the world and on what it means to be human.

The typical person, as Leonardo da Vinci noted, "looks without seeing, listens without hearing, touches without feeling, eats without tasting . . . [and] inhales without awareness of odor or fragrance." We are guilty of underappreciating—and underestimating—our sensory powers; after all, they circumscribe every waking moment. Observing how familiarity dulls our senses and anesthetizes us to the wonder of existence, the biologist Richard Dawkins suggested that "we can recapture that sense of having just tumbled out to life on a new world by looking at our own world in unfamiliar ways." Looking at our evolutionary family tree is one such way. We share a deep past with all creatures, but those I have chosen—from sea, land, and air—epitomize one or more of the various senses. The spookfish has an uncanny ability to detect light in the ocean's bathypelagic depths. The star-nosed mole navigates sunless subterranean tunnels through touch, whereas on moonless nights, the male giant peacock moth finds females miles away through smell. An exploration of such excesses proves there is more to unite than divide us. Our furred, finned, and feathered relatives offer insights across the range of human experience in all its shortfalls and surfeits. Through their eyes, ears, skins, tongues, and noses, our familiar and ordinary become unfamiliar and extraordinary, and curious new senses emerge.

The sensorium that we parrot from nursery—sight, smell, hearing, touch, and taste—was set out over two millennia ago in 350 BCE by Aristotle in *De Anima* (*On the Soul*). His concept of five senses persisted through Shakespeare's five wits and, to this day, remains a near-universal belief expressed across cultures, not only in everyday conversation but also in scientific literature. However, modern science has proved Aristotle wrong. Today a human "sixth sense"—once confined to the realms of pseudoscience with tales of telepathy or other extrasensory perceptions—is not simply scientific fact but has been joined by a seventh, an eighth, a ninth, and more. "We still are in the grip of an Aristotelian view of our senses," said the philosopher Barry Smith, "but if we ask neuroscientists, they say we have anywhere upwards of twenty-two." The neurobiologist Colin Blakemore confirmed this: "Modern cognitive neuroscience is challenging this understanding, instead of five we might have to count up to thirty-three senses, served by dedicated receptors." Aristotle's sensorium is proliferating.

Expert opinion differs on the final tally because there is as yet no consensus on how to define a sense. This shifts as scientists interrogate the substrate behind our various sensory systems. Some argue that it is folly even to try counting separate senses, as perception is about integrating information across them all, a fundamentally multisensory experience. We confuse the issue further in day-to-day conversation by invoking senses of loss and love, guilt or justice, art and music. While debate continues, what is not contested is that our eyes, ears, skin, tongue, and nose support more than one way of seeing, hearing, touching, tasting, and smelling—and that Aristotle failed to identify a host of other senses that toiled tirelessly beneath his awareness. Science has since shown that our eye senses not simply space but time. Some suspect it may even sense location much like a navigational compass. Our inner ear hears, and also senses whether we are balanced and keeps us on an even keel. Our tongue smells and our nose tastes, as do other bits of our body. Our nose might also detect airborne messages that don't even have a smell. A strange variant of touch exists within our muscles that grants knowledge of where our body is, allowing us to move with coordination and without thinking; another might inform the profound sense of our self. Unaware of their workings, like Aristotle, many remain ignorant of these senses. Yet these and more alchemize into sentience. In his final article for the *New York Times*, written a few months before his death, the man once described as the poet laureate of neurology, Oliver Sacks, bade farewell: "I cannot pretend I am without fear. But my predominant feeling is one of gratitude. . . . Above all, I have been a sentient being, a thinking animal, on this beautiful planet, and that in itself has been an enormous privilege and adventure." Open your eyes, ears, skin, tongue, nose, and more to the everyday miracle of being sentient.

1

The Peacock Mantis Shrimp and Our Sense of Color

T RUE TO ITS VARIOUS NAMES, THE PEACOCK, PAINTED, OR harlequin mantis shrimp is one of the most colorful creatures on the Great Barrier Reef. Neither shrimp nor mantis, *Odontodactylus scyllarus* is more akin to a diminutive lobster with a kaleidoscopic carapace of indigos, electric blues, and bottle greens. Yet this captivating countenance belies a somewhat irascible temperament. One spring day in 1998, at the Sea Life center in the English seaside town of Great Yarmouth, a particularly pugilistic specimen named Tyson astounded onlookers by smashing through the thick glass wall of his aquarium. "He was clawing and snapping. Nobody dared touch him," the manager told the national press. "All our visitors assume our sharks are the man-eating killers, but they are pussycats compared to Tyson. His power is incredible." Tyson was not the first to attempt such a jailbreak; these marine crustaceans, known as stomatopods, have developed quite a reputation among aquarists and scientists. Indeed, research has shown that the peacock mantis shrimp uses its club-like arms to pack a punch faster and more forceful than any heavyweight boxer.

One scientist at the University of California, Berkeley, made it her mission to understand the mantis shrimp strike, but only because she had run into problems with her original research plan. "I decided to take a break from trying to study their sound production and look instead at a behavior they perform regularly, without hesitation," explained Sheila Patek. "It was a classic example of how failure can open up new and unexpected directions." Her first challenge was to find a camera system fast enough. "Standard high-speed video cameras, that film at 1,000 frames per second, are too slow to capture the creature's strike. They only show a single frame of blur," she said. An opportunity arose to team up with a BBC film crew and use the latest high-speed technology for low light conditions. "Low light is the critical issue when filming these animals," because, "if it's too high, you fry them." The experiment was simple to set up: a peacock mantis shrimp, a sacrificial snail loosely tethered to a stick—"they are aggressive animals, happy to strike whatever is placed in front of them"—and, sure enough, they soon had a recording of a shell-splitting impact. They had filmed the punch at 5,000 frames per second, and, playing it back, they slowed it down by a factor of three hundred. "It was still pretty darn fast," Patek told me. "Even a back-of-the-envelope calculation for the speed and acceleration of the strike put them right at the outer limits of what people had ever seen." The final calculation was more surprising still: it was the fastest strike ever recorded in the animal kingdom. "It is a glorious moment as a scientist to see something for the first time and recognize how special it is," Patek added. The calcified club accelerated like a bullet in a gun, reaching its target in three-thousandths of a second, at velocities approaching 80 kilometers (50 miles) per hour. "But that was not the end of the story."

Patek decided to film the behavior at even faster speeds. "At 20,000 frames per second, we saw an incredible flash of light where the limb hit the snail, that then spread over the shell," she said. "I recognized it instantly." She was looking at a potent phenomenon called cavitation, which occurs where areas of water moving at vastly different speeds meet and the pressure drops. "This results in the water literally vaporizing and when that vapor bubble collapses, it does so with such destructive force that it emits sound, heat, and light." The experiments revealed that the force behind the peacock mantis shrimp's fist is so great that sparks really do fly. The knockout blow spells doom for aquarium walls and any snails unfortunate enough to be within reach. Patek's research enabled the Guinness World Records to claim it, relative to the animal's weight, as "the most powerful punch in the animal kingdom." But the mantis shrimp shows prowess beyond the boxing ring.

*

Just inland from the Great Barrier Reef and the Coral Sea, the University of Queensland's Brain Institute is perhaps an unlikely global hub for stomatopod science. "My major research love in life is the mantis shrimp," confessed Justin Marshall, the professor in charge of the Sensory Neurobiology Group. He and his team are often seen swapping lab coats for snorkels and scuba gear to brave encounters with these plucky crustacea and keep their aquarium well stocked. "The local fishermen call them thumb-splitters so we have to be careful," he informed me. "We collected this peacock mantis shrimp a few weeks ago on the reef just off Lizard Island. They are secretive creatures, often hidden away. This one was between a couple of rocks, so we put the net at one end, then prodded the other, and he shot straight into our trap." As Marshall peered down into a glass tank, a pair of protuberant, purple-hued eyes returned his gaze. "These eyes are unique," he explained. "Even the way they look at you is disturbing. A shrimp will fix you with its eyes, turn its back, scratch its behind, then turn back to eyeball you again, just like a monkey might do: as if they have primate-like awareness." Little seems to escape the stare of a mantis shrimp. Seemingly curious eyes swivel on stalks, independent of one another and rarely in the same direction or at the same time. Scientists have shown that whereas we humans need two eyes for depth perception, the mantis shrimp needs only one. This is the first of many visual talents. As Marshall told me, "Its eyes are more powerful than its right hook."

Marshall's fascination with Tyson and his brotherhood began some thirty-five years ago, on the other side of the world. He was beginning a PhD with Mike Land at the University of Sussex in England and had been scouting around for a subject when the decision was made for him by the visit of a foreign dignitary. "A larger-than-life African princess wearing a psychedelic caftan of many colors came to see the aquarium room," he recalled. "As she walked through the door, all the stomatopods leaped to the front of their tanks and waved their append-ages. I began to wonder if they might see color: a radical step for such a small-brained crustacean." Marshall decided to have a closer look. Under a light microscope, the surface of the peacock mantis shrimp's eye resolved into thousands of tightly packed hexagonal lenses called ommatidia, faceted like the compound eye of a fly. A line running horizontally across the middle caught Marshall's interest. "I could see a midband made up of six parallel lines of ommatidia, in which each ommatidium was bigger and more raised than those in the rest of the eye." To understand how these elements worked, he had to look closer still; he had to access their inner architecture.

With great care, Marshall froze, then finely sliced the midband and placed the sections beneath the microscope. What he saw through the eyepiece was extraordinary: "Each ommatidium was made up of light-sensing cells stacked one on top of another—three tiers in the first four rows, two tiers in the lower two." Yet their microstructure was not the most startling aspect: "I was expecting to see transparent things under the microscope, but lo and behold, I saw tiny blocks of bright, different colors instead." Reds, oranges, yellows, blues, pinks, and purples were scattered throughout the ommatidia—a rainbow hidden within the creature's eye. Similar colored oil droplets had been observed before in animals such as birds, where they filter light and enable color vision. "This was a pretty persuasive clue that these animals see color," Marshall told me. "I let out a litany of expletives and went to find Mike."

Marshall would need a rare piece of scientific equipment. "At the time there were only four of these machines in the world," he explained, "so Mike shipped me off to Baltimore for a few months, to Tom Cronin's lab. Tom had the kit and was the crustacean vision man." A microspectrophotometer passes a narrow beam of light through a microscopic section of cells, and by measuring what reaches the other side, it identifies what light they absorb. It would allow Marshall to examine the sensory cells within the mantis shrimp eye that receive and respond to light: its photoreceptors. The work must be done in near darkness, targeting photoreceptors mere thousandths of a millimeter across. Analyzing the ommatidia row by row, Marshall started to notice that the various cells within the rows absorbed different light wavelengths. In the first four rows, he found as many as eight types of photoreceptors, each tuned to a distinct color wavelength. Here was proof that the colored oil droplets he had seen were indeed filters and that the mantis shrimp's world was full of color. "These eight photoreceptors meant the peacock mantis shrimp has color sight more complex than any animal ever studied at the time and more complex than I could have dreamed up," said Marshall. "The story Justin brought back from America was amazing," agreed Land. "Some birds and butterflies may have as many as five, but *eight*!" Marshall took stock: "If this was shocking, there was more to come."

Marshall's investigations would uncover four additional photoreceptors for wavelengths of light that are invisible to our eye. Ultraviolet vision may not be unusual in the animal kingdom—it was known already in birds, bees, and butterflies—but it expanded the mantis shrimp's sense of color and brought the count of photoreceptors to twelve. "It was such a ludicrous excess of color capability that I was baffled. It did not make sense," admitted Marshall. Meanwhile, Land realized that "this was a color system quite unlike ours, or any other known animal." Further research revealed further excesses: eight more photoreceptors, including six for a property of light called polarization that specifies how it vibrates. Whereas color-blind octopuses see patterns of polarized light, the mantis shrimp detects not only color and regular polarized light but also circularly polarized light, which vibrates differently again. This last talent enables stomatopods to extract yet more information from the sun's rays. To our knowledge, no other animal can see circularly polarized light, so they use it among themselves as a secret channel of communication.

"The eyesight of these creatures is formidable," Marshall told me. "Four hundred million years ago, one of them got hold of an optics textbook, and now they are a physics lesson on a stick." When Tyson eyeballed the world beyond his tank, he did so with what Guinness World Records calls the "most complex eyes of any animal," with the "greatest color vision." No other eye approaches the shrimp's twenty different photoreceptors. According to Marshall, "We now know that the eyes of the mantis shrimp are out of this world." However, they also tell us something about the many ways that we humans see the world.

The amount of information conveyed within light is diverse, if not infinite. To take advantage of this, mantis shrimp eyes support many different ways of seeing; they sense ultraviolet light and regular and circularly polarized light, to name a few. Similarly, science can divide human sight into separate senses. As the Introduction to this book acknowledges, experts debate exactly how many, and the number they arrive at depends on how they define a sense. In *Great Myths of the Brain*, while contesting "the mistaken idea that we have precisely five senses," the cognitive neuroscientist Christian Jarrett suggested that if we are classifying according to photoreceptors, human vision can be subdivided into four senses, but if we are classifying according to visual experiences, the number is far greater. Despite our disparate eyes, one can argue that we share some of the shrimp's visual senses and that its greatest visual skill, a propensity for color, illuminates how we see rainbows. To understand the full range of human color vision, one must consider its antithesis: human color blindness—not the relatively commonplace incapacity to distinguish between red and green but the complete and utter loss of every shade under the sun.

*

In a far-flung swath of the South Pacific, north of Papua New Guinea, lies a cluster of Micronesian islands and the remote atoll of Pingelap. This one square mile fringed by beaches and coral reefs has a clear lagoon at its center, one main street, a school, and a congregation of churches. Pingelap is a picture-postcard paradise. Yet an inordinate number of its 250 or so inhabitants have been born with a rare impairment called achromatopsia: a condition that drains their vision of all color. They have never seen the sea-tinted skies, the sunshine yellow of the local teardrop butterflyfish, or the inflatable crimson neck of the great frigatebird as it booms out its mating call. Their world is confined to graduating grays and quickening shadows.

In 1994, the late neurologist Oliver Sacks embarked on a 12,000-kilometer pilgrimage from New York to what he called the "island of the color-blind." He had long been fascinated by complete color blindness since suffering an attack when he was a child during a particularly unpleasant migraine. Although it had lasted only minutes, it left an indelible impression. Sacks wrote, "This experience frightened me, but tantalized me too, and made me wonder what it would be like to live in a completely colorless world, not just for a few minutes, but permanently." Years later, he came upon achromatopsia on meeting the man he wrote about as Jonathan I., a painter who had lost his sense of color in a car crash. Jonathan I. had likened his condition to "viewing a black and white television screen." "My brown dog is dark gray. Tomato juice is black," he added; people appear "like animated gray statues," with "rat-colored" flesh, and everything is "molded in lead." His world had become impoverished, even grotesque. Although Jonathan I. could no longer remember or dream of color, Sacks wondered whether this was because he remained conscious of what might have been.

Pingelap could offer Sacks fresh insight because its color-blind islanders had been this way from birth. "I had a vision, only half fantastic, of an entire achromatopic culture," he mused, "where the sensorium, the imagination, took quite different forms from our own, and where 'color' was so totally devoid of referents or meaning that there were no color names, no color metaphors, no language to express it." He undertook the journey to Pingelap with Knut Nordby, a Norwegian vision scientist who, like the islanders, was born with achromatopsia. Stepping off the tiny prop plane onto Pingelap's concrete runway, Sacks and Nordby were greeted by groups of squinting children. Sacks realized it was the first time the islanders had met an achromatope from elsewhere and the first time Nordby had seen so many of his own kind. "It was an odd sort of encounter. Pale, Nordic Knut in his Western clothes, camera around his neck . . . [surrounded by the] achromatopic children of Pingelap—but intensely moving."

The team soon discovered that the color-blind islanders would go to lengths to avoid glaring sunlight. They would emerge from their homes only in the early mornings and evenings; many had taken to working as nocturnal fishermen. Those who did brave the daytime did so only with the protection of visors, wide-brimmed hats, and sunglasses. Achromatopsia is more than a simple absence of color. As Nordby explained, "I am easily dazzled and, in effect, blinded if exposed to bright light." Despite the inconveniences, the locals did not view color blindness negatively. Sacks learned the achromatopes hold a special place in local mythology as children of their god Isoahpahu. Nordby wrote, "Although I have acquired a thorough theoretical knowledge of the physics of colors and the physiology of the color receptor mechanisms, nothing of this can help me understand the true nature of colors." Yet he too found positives in his situation: "I have never experienced the 'dirty,' 'impure,' 'stained,' or 'washed out' colors reported by the artist Jonathan I.," and "I do not experience my world as colorless or in any sense incomplete." As Sacks watched Nordby taking photographs of the island, he was impressed by how little color blindness seemed to inhibit his sense of beauty. He wondered if Nordby saw "more clearly than the rest of us," whether to him the rich vegetation, which to us color-normals was a confusion of greens, was "a polyphony of bright-nesses, tonalities, shapes and textures." This thought exposed the gulf between his two encounters with achromatopsia; whereas Jonathan I. had viewed the condition as a blight, Nordby and the islanders seemed to appreciate its blessings. As another achromatope would later tell him, "We look, we feel, we smell, we *know*—we take everything into consid-eration, and you just take color!" Such an outlook begs the question whether, perversely, color might blind the rest of us to much of what the world can offer.

Presumably an achromatope's experience of reality could not differ more from that of the mantis shrimp. If there were a continuum for color vision, the achromatope's richly informed but monochrome view-point would be at one end and the crustacean's technicolor at the other. The points in between, where we fall, have dramatic effects on our own experience. In fact, there is sufficient variation between those of us with color-normal eyes to divide the public, as it did in February 2015, over whether an image of a dress was "blue and black" or "white and gold." Such diversity of visual experience is a compelling reminder that color is not out there in the world but within each of us. A centuries-old philosophical thought experiment runs, "If a tree falls in the forest, and no one is around to hear it, does it make a sound?" The American essayist, poet, and naturalist Diane Ackerman suggested something similar for color vision: "If no human eye is around to view it, is an apple really red?" The answer to both questions is no; color and sound do not exist without a spectator or a listener to see or hear them. Color is in the eye of the beholder. Ackerman added that the apple is also "not red in the way we mean red." The photoreceptors of our eyes register only a small bandwidth of the electromagnetic spectrum, which we call light. We perceive its various wavelengths as the rainbow's many hues. When sunlight hits an apple, the peel absorbs a portion and the rest is reflected, some of which enters our eye. We see the rejected wavelength only—the wavelength we perceive as red. As Ackerman puts it, "The apple is everything *but* red." Yet in an achromatope, something malfunctions that prevents that person from seeing red and any other color. To understand this inability is to clarify our own ability.

About thirty years before Sacks's visit, the prevalence of achromatopsia on Pingelap caught the attention of a young ophthalmologist from Honolulu University named Irene Hussels. Arriving on the island aboard the HS *Microglory* in 1969, she and her colleagues found this condition, which usually afflicts no more than one person in 30,000, affected as many as one in twenty of the islanders. Like many other small island communities, Pingelap's history is handed down by word of mouth. Talking to the tribal elders, they soon learned of a storm that had laid waste to the atoll some two centuries before. Further research revealed that over a few minutes in 1775, Typhoon Lengkieki had wiped out 90 percent of the inhabitants. The ensuing starvation killed more. Ultimately only twenty or so survived, including the king, Nahnmwarki Okonomwaun. Over the years that followed, the population began to bounce back, aided in no small part by his heroic breeding. Tellingly, Hussels learned that two of the king's six children with his first wife, Dokas, had been totally color blind; the rules of genetics dictate that for this to have happened, the king and his wife must both have been asymptomatic carriers of a gene for achromatopsia. Carefully tracing family trees, the scientists worked out that every living island achromatope was a descendant of Nahnmwarki Okonomwaun. The typhoon had sealed their fate when it spared the royal; it was his heroic inbreeding that bequeathed the dubious genetic inheritance.

Over the following three decades, Hussels married, becoming Maumenee, and although her research took her elsewhere, her thoughts remained on Pingelap. Then in 2000, working at the Johns Hopkins University School of Medicine, she was given the chance to lead a search for the royal gene responsible for color blindness. The geneticists took blood samples from thirty-two islanders—of whom half had the disorder—and compared their DNA. A previous study had highlighted the importance of a particular segment of chromosome 8, so Maumenee's team set about the arduous task of sifting through its more than a million nucleotides. Eventually they pinpointed a single mutation that, passed down from King Okonomwaun through the generations, was the cause of the islanders' achromatopsia. This mutation radically alters a gene that encodes a protein in the membranes of certain cells in the human eye known as cones, thereby ensuring the mass failure of all 5 million in our retina. Cones are the photoreceptors that grant us color; the microscopic marvels that open our eyes to rainbows, harlequins, and, aptly, the most conspicuous creature on the Great Barrier Reef.

*

Our eyes have more in common with Tyson's than first impressions might suggest. Close inspection reveals striking resemblances at the level of cells and proteins. The cones of our retina are akin to the color photoreceptors that Marshall found in the mantis shrimp midband. Moreover, we now know that both are saturated with the same class of light-responsive proteins known as opsins. When Charles Darwin wrote *On the Origin of Species* in 1859, he was bemused by the eye: "To suppose that the eye, with all its inimitable contrivances . . . could have been formed by natural selection, seems, I freely confess, absurd in the highest possible degree." He—and indeed, the scientific world at the time—was unacquainted with opsins. They have since been found in various guises in eyes everywhere—from corals to katydids, sea squirts to squirrels, as well as mantis shrimps to humankind—proof of life on earth's deep and shared history. Indeed, current molecular science dates the mother opsin to over 700 million years ago, not long after the common ancestor of all animals took its final breath. Opsins are the most studied of all sensory receptors. Cones may be the smoking guns of color, but opsins are the trigger.

Sight starts when photons of light—packages of energy so small they are point-like—enter the pupil of our eye, continue through the vitreous to the back of the eyeball, and reach the photoreceptors of the retina. Here they hit an opsin, setting in motion a cascade of chemical reactions that ends in an electrical spark. Light becomes a signal that shoots down nerves to the brain, and the external world becomes something we can perceive internally. Scientists still have little idea how nerve cells give rise to inner experience—how the tangible becomes intangible. Yet this astonishing transformation plays out with mundane, microsecond repetition. The various opsin structures fine-tune the eyes to different properties of light. Human cones are primed with one of three kinds: opsins sensitive to long-wavelength reds, medium-wavelength greens, and short-wavelength blues make, in turn, the red, green, and blue cones. As this trio reacts, in differing intensities and combinations, our brain compares their outputs to create the perception of color. Colored light does not mix like paints on a palette; combining all the colors of a rainbow creates not a sludgy mess but pure white. If red and green cones are activated, we perceive yellows and oranges, whereas differing combinations of green with blue cones can make teals and turquoises, and blue with red cones might make violets and indigos. When we succumb to Maumenee's mutation, our red cones cannot register light bouncing off the apple, our green cones cannot register light off its leafy branch, our blue cones cannot register light off the summer sky, and crucially there is no interaction between the three to conjure our world of color.

The calculus of color perception across the animal kingdom is relatively straightforward; species see different rainbows depending on how many different color receptors they have. Monochromats with just one type of cone—owl monkeys, seals, and whales—are color blind, so they see the world in one hundred shades of gray. Dichromats with two—a list that contains nearly all mammals, from anteaters to zebras—see a reduced rainbow. The dog, for example, has blue cones like ours, as well as another cued to wavelengths between red and green light, which is why it cannot distinguish a red ball from grass. Despite this, the vision scientist Jay Neitz calculates that because the second cone type offers each dilution of gray around one hundred new possibilities on the yellow-to-blue scale, dogs can see around 10,000 different hues. The addition of a third type of cone translates to a theoretical third dimension of color mixing to create color "space." We can see many subtle shades beyond the rainbow—walnuts, caramels, umbers, silvers, bronzes—but the thousands of words we have barely start to describe all we perceive. Individual variation combined with the subjectivity of experience makes a definitive count elusive. Neitz again calculates—as each of the 10,000 shades mixes with the one hundred discriminable steps from red to green—that our species can see at least a million different colors. Most vision experts agree that an average and unremarkable human eye more probably sees as many as several million. Either estimate is a giant leap along the color continuum from the achromatic experience of Knut Nordby. As rare mammalian trichromats—kept company only by the great apes, baboons, and macaques—our sight is far from ordinary, but it pales when compared to that of Tyson.

*

Justin Marshall was stunned by his discovery of the mantis shrimp's twelve color photoreceptors—four times our count—because he understood the theory of how they could combine. He saw how this pocket crustacean could detonate our understanding of color space. When experts are asked to describe what a mantis shrimp might see, their answers often invoke superlatives: its world has been called "the richest, most harmonious chorus of colors imaginable" and "a thermonuclear bomb of light and beauty." As Marshall put it, "We all start waving our arms around and resort to words like psychedelic or mind-bending if we think of these shrimps as potential dodecachromats." According to the way in which cones or color photoreceptors multiply, the shrimp's twelve could conceivably create a palette of one hundred hues to the power of twelve—one million million million million, or a septillion—a number with such an embarrassment of zeros it is ungraspable. "If our mind boggles at the potential for dodecachromatic color space, how on earth can a shrimp brain decode it?" Marshall asked. Such excess might be beyond our imagination. Yet a paper in an obscure scientific journal from 1948 hints at the existence of humans who might see something of the shrimp's worldview.

In the 1940s, the Dutch physicist Hessel de Vries was studying red-green color blindness. This condition is linked to the X chromosome (one of the two sex-determining chromosomes), so it is more likely to express itself in men than in women, rendering them as color compromised as their four-legged "best friend." It is often called Daltonism in memory of the first person, both scientist and sufferer, to write about it in 1794. Allegedly John Dalton realized he saw the world differently only when he unknowingly broke the all-black dress code of his Quaker meeting house in the English Lake District, arriving one morning in scarlet hose. Daltonism can result either from an absence of red or green cones, making a dichromat, or when all cones are present but tuned somewhat differently. This latter kind was first discovered in the late nineteenth century by the Nobel Prize–winning English physicist Lord Rayleigh (John William Strutt, third Baron Rayleigh). He observed that when color-blind individuals were asked to mix red and green lights to match a standard tone of yellow, some added more red than most, while others added more green. Rayleigh theorized that despite their Daltonism, they had three cones but the "red" cones of the first group must be somehow less sensitive to red; the "green" cones of the second group were similarly less sensitive to green. Such anomalous trichromats would later fascinate de Vries, who would submit them to further color tests. On one occasion, a man brought his two daughters along to the experiment, whom de Vries also tested. Although they exhibited none of their father's color blindness, the way they created color mixes also differed from the norm. Pondering this, he wondered whether these daughters might see more, not less, than others. Perhaps, in addition to the normal sweep of red, green, and blue cones, they had also inherited a fourth cone, their father's anomalous one. Conceivably such human tetrachromats could be capable of superhuman vision; a fourth cone could fracture our familiar colorscape into myriad, more subtle shades. When de Vries came to publish his latest experimental results in the August 1948 edition of *Physica*, his theory of the human tetrachromat was confined to a single sentence, buried on the last page. He never revisited the possibility, nor did anyone else until almost half a century later.

Around the time that Justin Marshall was deciding on the focus of his PhD dissertation, farther inland, at the University of Cambridge, another graduate was doing the same. Gabriele Jordan had grown intrigued by de Vries's long-forgotten paper: "It was not an easy read. De Vries had clearly been a very intelligent man—in fact he was nearly given a Nobel Prize. I would love to have met him," Jordan told me. She was struck by his throwaway line and its promise of the possibility of exceptional eyesight in humans. "I realized that the field had come a long way since his observation and we knew a lot more." The scientific landscape—vision research included—had been transformed by molecular genetics. While sequencing the DNA of the human retina's three cone opsins, a team from Stanford University had found that the red and green genes not only are next to one another on the X chromosome but also share 98 percent of their DNA.

This discovery exposed precisely how Daltonism can occur in trichromat males. "Highly similar genes are known to recombine in new ways," Jordan explained. "In the case of the red and green opsin genes, they mix to create hybrids, whose photopigments will have spectral sensitivities somewhere between normal red and green cones." Alongside a blue cone, males less sensitive to red would have a green cone and a hybrid more tuned to green, whereas those less sensitive to green would have a red cone and a hybrid more tuned to red. Crucially, the genetics also confirmed de Vries's notion of a human tetrachromat and revealed how frequently fourth cones arise in the population. "We now know that 6 percent of Caucasian males carry these hybrid genes and are, by definition, anomalous trichromats," said Jordan. Just as de Vries had seen, each could father a tetrachromat daughter; similarly, each could have a tetrachromat mother, "so 12 percent of women should carry these hybrid genes and have retinas with four classes of cone. A high enough percentage to get me thinking." Jordan picked up the baton from de Vries and embarked on a quest for the world's first known tetrachromat. These twelve women in one hundred would be unaware they possess a fourth cone cell or see the world differently. Jordan realized the best way to track them down would be via their color-blind sons. "A month, maybe two, was what I thought it might take me at the beginning." The odds appeared stacked in her favor, but she could not have known the challenges that lay ahead.

Finding the test subjects proved to be the easy part. Thirty-one women, with anomalous trichromat sons, volunteered. "These women, I knew, ought to have four types of cone on their retina," said Jordan. Next, she faced the far-from-trivial task of devising a visual test for something that was beyond her own perception. "Our whole world is tuned to the trichromat," she said. Not only are clothing dyes, paints, and printer inks manufactured by trichromats for trichromats but also all color monitors—from televisions to computers—work on the same color principle as the trichromatic eye, creating colors from the building blocks of red, blue, and green. "There were no off-the-shelf instruments I could use, so I had to engineer an entirely different colorimeter from scratch, one that could create and control for subtleties of color that I could not see." The design took months of careful experimentation in a darkroom, splitting beams of white light, filtering it through various combinations of lenses to distill fine, spectrally pure bands. "I knew that the spectral sensitivity of this fourth opsin would be between red and green, so I decided on a version of the Rayleigh match test." Much as Lord Rayleigh had done a century before, she started to test the subjects, but with a twist. Rather than asking them to mix red and green lights to match a pure yellow light, she incorporated a fourth light to investigate this extra dimension of color: she asked them to mix red and yellow lights to match an orange-and-green mixture. Quietly she hoped they would be hard-pressed to make many matches. "A normal trichromat would be able to make a whole range of matches, but a true tetrachromat would discriminate between all these mixtures except a single one."

Some of the women were not happy with their mixes, complaining, "I want to add more orange to the mixture, not red," or, "It's the wrong kind of orange. It looks rather pink when I add more red." Nonetheless, subject after subject, in test after test, made a whole range of matches. Only one woman exhibited vision that was other than ordinary, but even she fell short. Jordan's dreams of finding a functioning tetrachromat were fading. "The whole experiment had taken a year, and still the evidence was inconclusive. It seemed that a fourth cone cell did not guarantee superior color vision at all," recalled Jordan. "Color vision does not just depend on the number and type of cone opsins. To perceive colors and be able to discriminate between them, the inputs need to be compared." She proposed that the women tested so far were "weak" tetrachromats. "Only when the cortex gains access to the fourth signal will an individual perceive colors along a dimension denied to color-normal people." So Jordan's focus shifted to finding the world's first "strong" tetrachromat.

In 1999, Jordan moved to the Institute of Neuroscience at Newcastle University. "When I arrived, the lab was still being built. I needed to find grant money to buy the equipment. I bought so much I was given an optical table for free, but it was so heavy that it had to be winched into place, so I suppose we were lucky that the building didn't yet have a roof." Then the lab flooded, ruining key scientific equipment. "That's when I wondered whether God was trying to tell me to give up and find a new challenge." She persevered and set up the Tetrachromacy Project to source new subjects. Over the interim decade, she had fine-tuned the experimental setup and was now running a somewhat different test: "a discrimination version of Rayleigh's original matching task." Now when subjects were led into the darkroom, they were shown three different lights in quick succession and not asked to make matches but to identify the odd one out. Two of the lights were monochromatic yellow of differing brightness; the third was a varying red-green mixture that would appear discernibly different only to someone with strong tetrachromacy.

The first volunteer, a PhD student, seemed to pass the challenge with flying colors, but any excitement was short-lived. "She was so clever. She had heard the shutters releasing the beams of light and worked out that one click was for yellow; two clicks for the red-green mixture. It's amazing the cues people are alert to when asked to perform a sensory test." So Jordan asked her next subjects to don headphones as she streamed white noise to mask any other "unforeheard" clues. "We had been testing a group of about fifty women: thirty-one were mothers of anomalous trichromats, carriers of a fourth cone, but frustratingly one after the other failed the tetrachromat challenge." The women were performing no better than normal, oblivious to the many subtle spectral differences that flicked before them. Then, on the morning of April 20, 2007, a subject code-named cDa29 took the same tests, and with very different results. "At first, I couldn't believe it," Jordan told me. "Every test we put in front of her, she got right—no errors—and her responses were instant, without hesitation. She was making color discriminations with ease. We ran the trial again, then again, even a fourth time: still, zero mistakes. She was very unlike every subject I had seen before. It was compelling. After she had left, I was so excited that I jumped up and down!" Jordan was advised to test her again—"you can't be too careful"—and a month later watched yet another perfect performance unfold. "I like to think the discovery would have made Hessel de Vries smile," she told me.

An exercise that Jordan had initially thought would take months had taken some fifteen years. At last, she had identified the world's first "strong" tetrachromat, a doctor from the north of England who, until that day, had never realized her outlook on the world was anything special. "She had no idea, but there is no doubt that she is the real deal," said Jordan. "She occupies a perceptual dimension that is denied the rest of us." Perhaps the usual trichromatic world appears as bereft of color to her as John Dalton's would to most. "This private perception is what everybody is curious about. I would love to see through cDa29's eyes."

On the other side of the world, on the east coast of Australia, another woman has been identified with tetrachromacy. She has devoted much of her life to trying to share her vision. As a child, Concetta Antico was drawn to color and even decided on a career as an artist at the age of five. Yet she never suspected that the world she saw was different from everyone else's. "Growing up, you don't question what you see," Antico admitted. "It's only now, looking back, that I realize I was always different." Many years and a move to Los Angeles later, she learned about the condition of tetrachromacy from someone buying her paintings. "I was instantly fascinated. Who wouldn't be?" she asked. "The more I read about it, the more fascinated I became." Not long after, early in 2013, she walked into the Color Cognition Laboratory at the University of California, Irvine, having tested positive for a fourth cone cell. The cognitive scientist Kimberly Jameson knew instantly that Antico was unusual. She realized that "in addition to the genetic potential for tetrachromacy, Concetta has a considerable history of art training. She evaluates the uses of color and light, making hundreds of color space decisions every day." Jameson was particularly impressed by Antico's artworks of dimly lit scenes: "If you look at her pictures of dawn and dusk, she uses many colors." These essentially gray, monochromatic landscapes are portrayed in vibrant pastels; tree silhouettes are rendered in magentas and mauves, their shadows in madders and russets. Antico insists these spectral shades are not imagined. "The colors I paint into twilights are not artistic expression. Where you see grays, I see a rich and beautiful mosaic of lilacs, lavenders, violets, emeralds." She talks as if colors splinter beneath her gaze. "Take what you call white. You might see lead whites, ivories, chalks, silvers, warm whites, cold whites, but I see so many more subtle shades, most without a name." Jameson submitted Antico to a battery of visual tests and found her subject perceives far finer gradations of color than most others. The scientist argues the artist not only has a fourth cone but also exemplifies the power of practice. According to Jameson, "Concetta is the perfect storm for tetrachromacy": the embodiment of a spectacular synergy between nature and nurture.

Just before learning of her exceptional gift, Antico received news of a different kind. Her eight-year-old daughter was diagnosed with a form of color blindness less common than Daltonism. Tetrachromat mother and rare dichromat child are both unusual, with views of the world that are diametrically different. Antico has resumed painting with renewed fervor. She has returned to Australia and established a gallery in Byron Bay. "I'm working furiously right now. This lockdown from COVID has been a blessing in disguise." She uses paint in the hope of communicating her vision. "I wish that my daughter might see a small part of what is available to me. In fact, I wish everyone could realize how beautiful the world actually is; then perhaps they might value it more." Antico's attempts are in vain; she has set herself an unattainable goal. Her paintings represent a world beyond our reach and remind us that we cannot see through another's eyes. This unbending reality also bars us from the perceptual worlds of achromatopes, from Jonathan I.'s agonies of rat-colored flesh or Knut Nordby's hopes to know more intimately the quality of things. Gabriele Jordan first grappled with it when designing a test for colors beyond her ken—then, when faced with her long-awaited quarry, knowing she was excluded from cDa29's vision of the world.

*

The eye of the peacock mantis shrimp remains unrivaled; we know of no other creature with such optical panache. If the achromatopes of Pingelap, with their cone-killing gene, are living proof that color comes from within, then Tyson shows how our sense of color is consistently underestimated. Recently, however, Justin Marshall's Queensland laboratory uncovered a twist in the tale of the shrimp. Rather than focus on the workings of its eye, researchers tried to explore its private perceptual world—what a shrimp might actually see. Taking advantage of their proclivities for fighting and fine dining, Hanne Thoen trained the crustacea to approach, nibble, and sometimes strike a fiber-optic cable colored, say red, in exchange for mouthfuls of juicy crabmeat. Next, she presented her subjects with a choice between a red and an orange fiber optic, rewarding the shrimp only if it stuck to red. Then she adjusted the orange cable—minute tone by minute tone, through tawny, brick red, crimson—until its wavelength approached the red one. "Thoen had to dissect the whole spectrum and train shrimps to attack its component colors for over two years. The experiments drove her nuts," Marshall told me, but the results were revolutionary. "I could not believe the data at first, except her method was rock solid. There is no doubt the shrimps performed appallingly, unable to discriminate between colors we can see easily." Whereas we identify shades barely a nanometer apart, they lose the ability with colors less than twenty-five nanometers apart.

Dreams of "thermonuclear bombs of light and beauty" fizzle. Marshall now believes the mantis shrimp displays some of the *worst* color vision in the animal kingdom. "Whereas our color comes from the comparison of neural signals from three cones," he explained, "the shrimp must use the signals from its twelve photoreceptors to perceive color in a fundamentally different way. Over 400 million years of independent evolution and, yet again, they have arrived at a solution different from every known animal." The team remains ignorant as to how information from the shrimp's sensors combines in its brain to create the perception of color. Perhaps the shrimp does not compare and discriminate between hues but more simply recognizes them. Perhaps this takes place more quickly. Perhaps, alongside legendary lightning punches, the shrimp also boasts the fastest sight on the planet.

These are questions for another day, for another painstaking and patient experimenter. Meanwhile, it would seem that despite having a mere quarter of the shrimp's color receptors, our brains compensate for our eyes' shortcomings, allowing us to perceive more color in the world than its recently deposed king. Take a moment. Look up from this page and open your eyes to your surroundings. Let the peacock mantis shrimp reveal that the way we see the world is, quite simply, superlative.

2

The Spookfish and Our Dark Vision

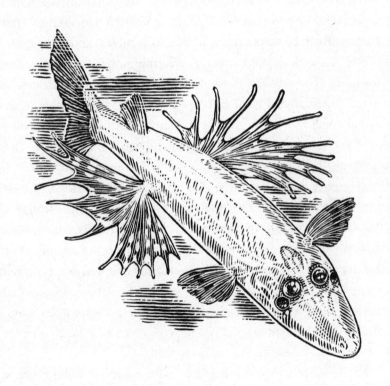

ON A CALM AND CLOUDLESS DAY IN JULY 2007, THE GERman research ship FS *Sonne* left Apia Harbor in Samoa, bound for the archipelago of Tonga and the deepest seas in the Southern Hemisphere. Onboard were scientists from around the world, including two British biologists, Ron Douglas and Julian Partridge. "It was a rare chance to take part in a voyage of discovery to bring back a clearer picture of life in the deep than had been seen before," recalled Douglas. "Everybody knows that 70 percent of the planet is covered in ocean, but we forget its depth. Its third dimension means it makes up 99.9 percent of the habitable planet." Beneath this stretch of Pacific Ocean lies topography more spectacular than anything above sea level. Tectonic plates converge, creating mountain ranges and ravines more colossal than those on land. The Tonga Trench has been mapped to just over 10,800 meters (35,400 feet) deep, second only to the Challenger Deep in the Mariana Trench. "You could quite easily put Mount Everest in one of these trenches and you wouldn't see it," Douglas told me, "but more people have gone to the moon than have reached 10,000 meters beneath the sea." For him and Partridge, the deep ocean—cold, black, and seemingly boundless—is the final frontier. "It is the least understood environment on earth," Partridge added, "and so completely out of our experience that it is difficult to imagine." It lures scientists with its promise of alien life-forms and new ways of seeing. Some seventy years before the *Sonne* would set sail, two pioneers defied death and plunged into this *aqua incognita* to experience it firsthand, in a contraption that had originally been conceived as a sketch on a scrap of paper by President Theodore Roosevelt.

The Bathysphere was the most rudimentary of submersibles: a cast-iron sphere only 1.5 meters (5 feet) across, weighing some 2,000 kilograms (4,400 pounds). On August 15, 1934, off Nonsuch Island in Bermuda, it would bear William Beebe and Otis Barton six times deeper than any human had ventured before. Onboard an open-decked barge called the *Ready*, the men had squeezed headfirst through a small opening in the metal casket and, cramped knee to knee, been sealed in from the outside by a door hammered into place with ten large bolts. The Bathysphere was tethered to the mother ship by a steel cable, providing electricity and a phone line to the surface, as well as saving it from sinking without a trace. This lifeline spooled out as the winch lowered the explorers over the side, under the waves, and down. Each man sat tight and gazed through one of the two portholes into the fathomless void. Thick fused quartz panes, made from melted sand, combined strength with transparency to all light's visible wavelengths. "I sat crouched with mouth and nose wrapped in a handkerchief to prevent condensation, and my forehead pressed close to the cold glass—that transparent bit of mother earth which so sturdily held back nine tons of water from my face," wrote Beebe in *Half Mile Down*. "I felt as if some astonishing discovery lay just beyond the power of my eyes." Yet a novel experience was already playing out before their eyes. Beebe and Barton were the first to witness the physics of what happens to sunlight as it journeys down and to understand that the deep is not the realm of perpetual darkness often imagined.

As the Bathysphere sank, the voyagers were struck by how the quality of light changed. They saw visible light's rainbow spectrum gradually being whittled away, hue by hue. "The first plunge erases, to the eye, all the comforting, warm rays of the spectrum," recalled Beebe. The color red faded within the first 15 meters (50 feet), then vanished. "I happened to glance at a large deep-sea prawn," and "to my astonishment, it was no longer scarlet but a deep velvety black." The ocean absorbs colored lights at different depths, according to their wavelength. Red, the longest wavelength, goes first, followed by the somewhat shorter orange. At 50 meters (160 feet), Beebe saw yellow, its shorter-still neighbor, disappear. "Yellow is swallowed up in the green. We cherish all these on the surface of the earth and when they are winnowed out . . . [the remaining spectrum] belongs to chill and night and death." The greens bled imperceptibly into the blues until, at around 200 meters (650 feet), "The last hint of blue tapers into nameless grey [as] the sun is defeated and color has gone forever." The men had dropped beneath the zone of color and daylight into one of grays. Above the waves, twilight ends the day, but in this watery part of the world, it is the day. Photons, nature's fundamental particles of light, followed the Bathysphere down, but fewer and fewer of them; the light intensity declined by about one and a half orders of magnitude for every 100 meters (330 feet). Twilight turned to night.

The explorers were now making use of a different sense from the one that had enabled them to see at the bright, sun-flecked surface. Scientists call it scotopic, as opposed to photopic, vision. Derived from the Greek words *skótos* (dark) and *opia* (vision), scotopia gives night sight rather than daytime color sight. We use this, our eyes' second sense, to see under starlit skies. It enabled the divers to peer through the gloaming as they dropped into an endless night. At 600 meters (nearly 2,000 feet), Beebe noted how this too failed, and blackness closed in on them, "The sun, source of all light and heat, had been left behind." Eventually their descent slowed, then stopped. The Bathysphere had reached 923 meters, over half a mile under the sea. Beebe later wrote:

> There came to me at that instant a tremendous wave of emotion, a real appreciation of what was momentarily almost superhuman, cosmic, of the whole situation; our barge slowly rolling high overhead in the blazing sunlight, like the merest chip in the midst of ocean, the long cobweb of cable leading down through the spectrum to our lonely sphere, where, sealed tight, two conscious human beings sat and peered into the abyssal darkness as we dangled in mid-water, isolated as a lost planet in outermost space.

The divers were like earthbound astronauts and, quoting British biologist Herbert Spencer, Beebe claimed he "felt like 'an infinitesimal atom floating in illimitable space.'" The oceanic abyss is darker than a night without either moon or stars. It is perhaps the most lightless place on our planet. Yet it teems with unblinking, nonhuman eyes.

✳

Volumetric analysis on the brains of deep-sea fish reveals that for the vast majority, sight is the most important sense. Many fish generate their own light in a biological firework display called bioluminescence. The lanternfish creates beams that sweep the sea like headlamps. The dragonfish produces wavelengths that only it can see, leaving its victims blissfully unaware of their impending fate. In contrast, the anglerfish hopes its prey will notice and be lured toward its rod-like bioluminescent barbel; its fierce jaws stay hidden in the shadows. Bioluminescence is also used to foil predators. A species from the spookfish family, *Opisthoproctus soleatus*, relies on a bellyful of symbiotic, luminous bacteria to save it from becoming a meal. It uses the same concept developed by the US Navy during World War II to camouflage bomber aircraft. Just as Project Yehudi designed planes with under-wing spotlights, the fish's glowing belly camouflages its silhouette against sunlight to hide it from watching eyes below. In this fish-eat-fish world, survival is a game of hide-and-seek that prioritizes the sense of sight. Photons from bioluminescence may be sparse, but a few others from the sun penetrate the deep, reaching as far down as a thousand meters (3,330 feet). These glimmerings may have proved too faint for Barton and Beebe, but not for those on the other side of the Bathysphere's protective shell.

Over the past century, scientists have plumbed the ocean and hauled up more and more extraordinary eyes. The largest known belongs to the giant squid *Architeuthis dux*; it is the size of a dinner plate. Those of the cockeyed squid are, as its name suggests, of unequal size, with the larger one pointing upward to capture what is left of the sunlight. Put simply, big eyes gather more light than small ones. The ocean's finned inhabitants have similarly peculiar optics. "We are learning that deep-sea fish have the best dark vision of all vertebrates, on the land or in the sea," said Julian Partridge. "The oceans may be among the most photon-restricted environments on the planet, but these fish have evolved various techniques to make best use of the photons that are there."

Douglas and Partridge have come eye-to-eye with more deep-sea fish than most others. They understand how these creatures push vision to its limit. "When you've only got so much space in your head for an eye but you need a big pupil, the compromise is to trim off the sides of your eye, making a tubular shape, giving a very narrow field of view but a very bright image," Partridge explained. This tubular design is typical of deep-sea fish such as spookfish, including the glowing-gut *Opisthoproctus*. "Their eyes look like telescopes stuck on the side of their head, pointing up toward the surface and the light." Deep-sea fish also possess a layer of crystals beneath their retina; this tapetum reflects any photons that failed to initially hit the retina back into the eye. According to Douglas, half the light that reaches the retina is not absorbed and passes straight through. "Humans have a layer of melanin backing our retina to absorb such photons—we don't want light bouncing around, ruining the image quality—but deep-sea fish have to grab every photon they can get so this shiny tapetum gives them another chance at the retina." Footage taken at the start of the new millennium, in the ocean's twilight zone, revealed another spookfish, *Macropinna microstoma*, with yet another light-capture technique. The few preserved specimens that already existed displayed the typical upward-pointing, tube-shaped eyes, but the film showed that these were embedded within a bulbous and transparent head, which must have ruptured when previously fetched up from the depths. Large pupils, telescopic eyes, tapeta, and translucent heads are just some of the traits that facilitate dark vision. Douglas and Partridge hoped their *Sonne* expedition might bring even more to the surface.

✳

Once the ship had gained deep water, the search began. "The deep sea is indescribably vast, so you never know what you are going to catch," Douglas told me. Unlike Beebe and Barton, the researchers were not planning to get their toes wet. They started to trawl for specimens by releasing a Tucker net into the twilight zone. "Our methods for sampling this environment are very crude," Douglas explained. "Dragging a net the size of a soccer goal blindly through the vast ocean is like hunting through an Olympic swimming pool with a thimble." Also, they lowered a remotely operated lander overboard: a simple contraption made of scaffold that housed cameras. Untethered, unmanned, it started to sink even deeper, 10,000 meters (6 miles) to the ocean floor. "We baited it with tuna bought in Samoa, stuffed into some ladies' tights." A camera had never been sent so deep. Partridge recalled, "It's no trivial matter deploying unmanned cameras down to these depths." Douglas said, "It's relatively easy to deploy them; it's getting them back that's the problem." And when asked what the camera might find, he added, "What do you expect to see when you switch on the light down there and make a hell of a noise? There is a joke that it will be only those too old, blind, or too stupid to move." Two weeks after they had set sail, on a Saturday afternoon, a net trawling at 700 meters (2,300 feet) was reeled in. Within its mesh lay a creature that was neither old nor stupid and far from blind. Indeed, it possessed a flair for vision that would leave the rest floundering in the dark.

"It was the last thing left in the bucket after all the other scientists had had their feeding frenzy," said Douglas. At first glance, the fish looked unexceptional. They laid it belly-down in the ship's laboratory and a colleague took some photographs. "We could tell from its body shape and eyes that it was an Opisthoproctid or a spookfish. But none of us had ever seen this particular species before." It displayed the typical tubular eyes to maximize pupil size, which lit up when the camera flashed, proving the presence of tapeta. The researchers flipped the fish over to photograph its underside. As the camera flashed again, it revealed something unexpected: another pair of reflections, hinting at two further tapeta. "Without those photographs, we might never have realized that this fish was interesting," Douglas admitted. "It looked like it had four eyes, and vertebrates with four eyes don't exist." The team appeared to have hooked a four-eyed spookfish.

Scientific literature would later reveal that the creature had been described once before, in 1973, by a biologist trawling through ichthyology collections in a Copenhagen museum. "To my knowledge, Randi Dilling Frederiksen never even went to sea," said Douglas. "His material is likely decades old, from the *Dana* expeditions of the 1920s, given by Ole Munk, one of the godfathers of deep-sea fish eyes." It meant that "Four Eyes" already had a rather grand name—*Dolichopteryx longipes*, from the Greek *doliochós* (long) and *ptéryx* (wing), for its large, delicate pectoral fins—so the team called it Doli for short. However, its special gaze had been entirely overlooked, so it was with great anticipation and no small amount of care that they shipped their strange spookfish to a colleague in Germany for a closer look.

As professor of anatomy at Tübingen University nearing retirement, Jochen Wagner has spent many hours in darkened rooms, peering down microscopes. "If you want to work with light, you need to control it and that means you have to be in the dark," he said. When Doli arrived, he embedded its eyes in plastic so he could then slice them into microscopically fine sections, barely a couple of microns, or a few thousandths of a millimeter thick. With the lights dimmed, he took his first look. He saw that the secondary eyes were clearly joined to the main eyes; strictly speaking, the four eyes were two. "Sectioning showed that each eye consists of two parts separated by a dividing septum," explained Wagner. "There is the main, cylindrical, tubular eye that points upward and a smaller ovoid outgrowth from its side, a diverticulum, less than half the size, that points downward." Then he directed his gaze toward the two sets of optics. The upper main eye appeared like that of most Opisthoproctids, but the lower pseudo-eye lacked a lens. He looked closer. "One particular area would not stain and looked really fuzzy. I couldn't focus on it, which was very odd, but given I like to play around with specimens, I didn't think too much." He tried dark-field illumination, so that particular area was lit indirectly. "Curiously, this layer clearly showed it could refract light." Finally, he turned to dark-field illumination with polarized light. "As I was slowly rotating the slide, this funny structure lit up like mad, then went dark again, and that's when I realized it was optically active." The mysterious structure was made of well-ordered crystals in parallel stacks. "There was no doubt about it: I was looking at a mirror."

Partridge then ran computer models with astonishing results. The mirror was perfectly curved to reflect light as a well-focused image over the pseudo-eye's retina. "A mirror was the last thing I was expecting to see forming images in the eye," Wagner told me. On hearing the news, Douglas recalled, "It was a stunning moment. We already knew Doli was special, but not that special"—to which Partridge added, "It is the only vertebrate to use a mirror to produce an image in nearly 500 million years of vertebrate evolution, including the many thousands of species living and dead." When they came to publish their discovery in January 2009, tales of the inimitable four-eyed spook hit newspapers around the world. Two years later, a remote-operated camera happened upon a living specimen and brought back the first footage of the fish in its natural habitat.

At 700 meters (2,300 feet), 100 meters (328 feet) beneath the point at which the light ran out for Beebe and Barton, the black ocean is scattered with sparse pinpricks of bioluminescence that burn bright, then fade like distant, dim stars. A translucent and slender spookfish glides through the void, still but for a gentle, propulsive ripple through its wing-like fins. Its upward-pointing eyes, with large pupils and tapeta, collect the scarce solar photons that have penetrated these depths. Its downward-pointing, mirrored pseudo-eyes gather photons from below: perhaps from the sun bouncing up off a neighbor, perhaps from another's bioluminescent signal. The two sets of optics radically expand the spookfish's light-capturing horizons, maximizing the number of photons that reach its four retinas, bestowing sight. The spookfish can also shed light on us. It seems that Beebe and Barton underestimated their, and our, capacity for dark vision.

＊

As humans, we assume our night sight is a poor second compared to that of owls, foxes, and deep-sea fish, but few of us have put it to the test. "Nowadays people rarely experience pure scotopic vision, because we don't spend time in very dark places," Andrew Stockman told me. We shrink from the shadows. We hold them at bay with streetlamps; we buy night-lights to console fearful children. Satellite imagery of the moonlit parts of our planet twinkling with electric lights may look mesmerizing, but it reveals the extent to which civilization has banished darkness. Stockman, a professor at University College London's Institute of Ophthalmology, has spent his career investigating human vision. In 1992, a research collaboration took him to the medieval city of Freiburg, Germany. After days in the laboratory, he would sometimes climb the surrounding hills and, as the sun slipped below the horizon, set off on a jog into the darkening Black Forest.

On nights with a full moon, light levels are about 1 million times dimmer than at midday. On moonless but clear, starlit nights, they are 100 million times dimmer, and beneath a thick canopy of trees, they become a hundred times darker still. Stockman was experiencing conditions as scotopic as parts of the deep ocean: "I would start running in the pitch black, but slowly and surely my eyes would start to adjust. It takes about forty minutes for our eyes to wholly adapt to the dark, but even before then, I was impressed by how much I could see." Once his eyes were fully adapted and true scotopic vision had kicked in, he was again surprised despite his familiarity with the science. "Everything appeared in monochrome, without detail, yet oddly light." His eyes were able to see in light conditions a billion times dimmer than daylight, but he also observed how his night sight worked slightly differently from normal. "Again I was expecting it, but it was striking nonetheless," Stockman recalled. "If I wanted to see something in the gloom, I had to fix my gaze just to the side of it, and then it would become visible." The ancient astronomers were the first to observe this idiosyncrasy. They learned that faraway stars, too faint to see when focused on directly, could, however, be seen askance. Our talent for dark vision lies in the cells of our eye.

Our retinas are the most light-primed surfaces in our body. Under magnification, each resolves into a mosaic of two types of light-sensing, or photoreceptive, cells. They are named for their markedly different shapes. The larger, bulbous ones tend to be the cone cells of the previous chapter. Although there are around 5 million on each retina, they are uncommon when compared to their slender, elongated neighbors.

These rod cells outnumber cones by twenty to one, reaching around 100 million in total. Stockman's extra sensitivity around the edge of his vision can be explained by the fact that cones and rods are not distributed evenly across the retina. No more than 1 millimeter across, the circular, central fovea is crammed with, and dominated by, cones, exclusively so at its core. This translates into the brightly colored and sharply focused sweet spot at the center of our visual field, on which we rely regularly, particularly when reading the tightly packed words on a page. Moving outward, cones are gradually replaced by rods. Then, at 2.5 millimeters, in an area encircling the fovea called the parafovea, rods start to outnumber cones. By 5 millimeters, rods reach their peak density, with as many as 170,000 per square millimeter. Like the early stargazers, Stockman had to avert his gaze—against the eye's natural propensity to fixate with its fovea—to ensure any attenuated rays would hit these rod-rich parafoveal areas and give rise to perception.

If the eye is a camera, then cones work like analogue color film, capturing glorious Technicolor in high definition, whereas rods work like black-and-white film, exposing panoramas drained of color and with lower resolution. Cones perform best in daylight, whereas rods excel when the lights go down. As Stockman ran into the thick-canopied Black Forest, his cones failed but his rods took over, so his path through the trees would have appeared like old movie footage, monochrome and somewhat indistinct. If cones are the microscopic marvels that combine to give us rainbows, then rods grant us sight on the blackest nights and give us the farthest reaches of our galaxy with its distant stars. The way in which they summon light from darkness is reflected in the eye of the spookfish.

Under a microscope, the retinas of deep-sea fish also resolve into a tapestry of tightly packed photosensitive cells. In contrast to ours, where rods jostle for position with cones, the fish retinas are made up entirely of rods. To Jochen Wagner, the structure of Doli's retinal rods appeared nearly identical to our own: "The dimensions of the rods' inner and outer segments are surprisingly similar in Doli and humans." Yet the single layer of photoreceptors that exists in our retinas appeared to be multiplied: "The center of Doli's main retinas is around 600 microns thick. It accommodates seven to eight separate layers of rods. It thins down in the peripheral areas, but still with three to four layers." Similarly, its accessory retina has up to five tiers. This packing and stacking of photoreceptive rods across all four retinas soon adds up, as Ron Douglas explained: "In Doli it makes up about half the depth of the various retinas, whereas in humans it makes up just a fifth." Consequently, Doli's rod densities are almost double ours—an inordinate amount of low-light-sensing technology that translates into unparalleled dark vision. Nonetheless, the rods we have are strikingly potent, as certain humans demonstrate. These people, like the spookfish, occupy the shadows.

*

Knut Nordby was the achromatope who accompanied Oliver Sacks on the expedition to the island of the color blind. As with many of the islanders, the color-sensing cones in his retinas did not work, but color blindness was the least vexing aspect of his condition. It also forced him out of the light. He wrote in unflinching detail about his experience: "As far back as I can remember, I have always avoided bright light and direct sunlight as much as possible." When he was growing up, Nordby's well-meaning parents tried to encourage him outside into the Norwegian summer sun: "My whole childhood was, in fact, a continuous struggle against the prevailing views of what is proper for children. . . . I preferred playing indoors with the curtains drawn, in cellars, attics and barns or outdoors when it was overcast, in the evenings, or at night." He concluded, "It is very clear to me today that the most debilitating, handicapping and distressing consequence of the achromatopsia has been my hypersensitivity to light. . . . [This] is usually referred to as photophobia but alas is nothing to do with irrational psychodynamic 'phobias.' In fact, I really like being out in the warm sun." The problem Nordby faced was that without sunglasses or a visor, his eyes would struggle to regulate incoming light. An involuntary squint soon turned into paroxysms of blinking. "The blinking frequency is slow at first, only once every four to five seconds, but the periodicity increases with increasing light intensity, to three to four blinks a second," he explained. Like deep-sea fish, the only functional photosensitive cells in Nordby's retina were rods. As these reacted, his system soon reached saturation, leaving him blinded: "If I open my eyes fully for more than one or two seconds [in bright light], the scene I am gazing at is quickly washed out, turned into a bright haze and all structured vision is lost. It can be very distressing for me, and sometimes even painful." He would then need to sit in darkness to allow his rods to reset so he could see again. Nordby shunned daylight because even on the gloomiest days, his rods were swift to react to the smallest amounts of light. His blight is our blessing: the same sensitivity that plagued him grants us the ability to see in the darkest places.

"Our rod, or scotopic, visual system is around a thousand times more photosensitive than our cone, or phototypic, system," said Andrew Stockman. The reason is in part due to the pigment that primes them to light: rhodopsin, an opsin from the same class of photosensitive proteins found in our cone cells. It is present in the rods of all animals that have backbones. "The rods of Doli and humans are both packed with rhodopsin," said Douglas. "The opsin protein, with its tiny chromophore of retinal, enables the first step of scotopic vision." Multiple rhodopsin molecules thread in and out of the rod cell membranes, each holding a light-reactive chromophore in a snug embrace at the rod surface. Rods not only contain more visual pigment than cones, but also their rhodopsin is more receptive to light than the red, green, and blue opsin cousins. When a photon of light hits its retinal chromophore, it responds by changing shape. "Rhodopsin's absorption of a photon is the primary event of vision in both us and deep-sea fish," explained Douglas. It summons the seemingly miraculous process of phototransduction, transforming dim glimmers from the outside world into electrical signals shooting along our optic nerve. "The transduction process is far more sensitive in rods than in cones, so rods can reliably respond to single photon absorptions," said Stockman. Rhodopsin and rods enable our eyes to react to the individual quantum particles of the universe.

A closer look at Nordby reveals how such sensitivity begets conscious sensation.

"Knut was not only a vision scientist but also a willing subject who traveled to laboratories around the world, including Ted Sharpe's one in Freiburg where I was working at the time," Stockman told me. "His rare condition helped us understand normal visual perception by measuring rod-only responses without any cone involvement." Researchers had become intrigued by a perplexing perceptual anomaly that occurs when we watch a flickering light. Normally, increasing the intensity makes the flicker brighter and more conspicuous, but if the light is flickering at fifteen times a second, then doing so prompts something strange. By turning up the dial, the flicker seems, as if by magic, to vanish and leave a steady, uninterrupted illumination. Stockman and Sharpe turned to Nordby. They presented him with the same flickering light and slowly boosted its brightness. His eyes behaved as theirs had: he too saw the flickering cease. This meant that the anomaly was in the rod system.

Scientists suspected that after a rod absorbs a photon of light, it sends signals down more than one set of nerves to the brain. Nordby provided proof. According to Stockman, "Nordby's results could be explained by the rods having two neural pathways that work at slightly different speeds." Crucially, these pathways merge before leaving the eye, with remarkable results, because one is traveling more slowly than the other. "The delay between the slow and fast rod pathway is between 30 and 35 milliseconds. This means that the signals are half a wave cycle apart, so on combining, they interfere destructively and cancel the flicker signal." The light still flickers in the room; its photons still hit our retina, and some are absorbed by our rods, the nerves still fire, but before their signals reach the brain, the flicker is erased. We fail to see what is actually before us.

"Our idea of two rod pathways was originally conceptual. The results we got from the experiment we did with Nordby have since been supported by a growing body of evidence," Stockman explained. We now know that the two pathways optimize our rod vision to different lighting conditions. Like the speed of camera film, they shorten or lengthen the time involved in creating imagery. The fast pathway, like fast color film, performs best in bright light and gives us photopic vision. The slow pathway, like slow black-and-white film, works in low light and gives us scotopic vision. Essentially the sluggishness allows more time for photon absorptions to accumulate, thereby amplifying the signal sent to the brain. When Stockman left the lab that evening for his regular run into the Black Forest, his rhodopsin-primed rods would have absorbed the few photons that filtered through the tree canopy, alchemizing light into electrical nerve impulses. His slow rod pathway would then have ensured these signals coalesced to create a blurred—but surprisingly bright—perception of the trail ahead. Our eyes are astonishingly sensitive to light. Quite how little we need for sight was not fully appreciated until recently.

✻

Alipasha Vaziri is a quantum physicist who also studies neuroscience and cellular biology. "The ultimate reason for why I do what I do is—like any other scientist—curiosity," Vaziri told me. "I became interested in biological questions about fifteen years ago. I realized as a physicist I could bring in new conceptual approaches and devise new technologies to tackle big unsolved problems." One such problem was the lack of progress on ascertaining the absolute threshold for human vision: the bare minimum of light that we require to see. The last notable advance had been in 1942, when American physiologist Selig Hecht and colleagues published a landmark paper, since cited over a thousand times; it established that humans are able to perceive light signals made with as few as five photons. "Hecht's fascinating discovery captured my imagination," admitted Vaziri, "but no one knew quite how few photons are needed for perception." The reason was the lack of a machine that emits photons with precision. "With traditional means, one can only control the average number of photons, but never the exact number of photons," he explained, "a consequence of the quantum mechanical statistics that photons have to obey." Vaziri began to wonder if there might be another way to approach the problem.

In 2016, his team at the University of Vienna set to work designing a new quantum device. They decided on a crystal of beta-barium borate and an optical process—with the technical name "spontaneous parametric down-conversion"—to split photon beams into photon pairs. By directing one stream of the split beam at a detecting device and the other at their subject's eye, the researchers realized it would negate the need for a machine that produces a regular number of photons. "This setup meant we could eliminate the variability and uncertainty issue of classical light sources, because when we got a signal at the detector, we knew for sure that a single photon was about to arrive at the subject's eye." It was a simple yet elegant solution. Next, they needed a light-proof environment. "That too was a challenge," Vaziri recalled. "When you're testing for the perception of one photon, you have to be totally sure there are no others present." They custom built a tiny chamber about the size of a phone booth, with highly engineered, photon-proof seams—between the walls, floor, ceiling, and doorway. "Even the ventilation joinery was folded and overlapped many times to avoid light leakage." Assembled within their optical laboratory, this darkroom within a darkened room was tested with a photon detector to confirm it was indeed quite lightless. Shutting the door was much like being plunged into the spookfish's watery abyss. "It's a kind of darkness that one does not typically experience," said Vaziri. "We are used to seeing at least a little bit after sufficient dark adaptation. Not seeing anything, whichever way you tip your head, whether your eyelids are open or closed, is slightly disorienting." The design phase had taken them a year and a half. They were ready to start the experiments.

Vaziri knew that the best volunteers would be those who were intellectually invested in the work: his students and postdocs. So the first experimenter-turned-subject entered the room and, like all who would follow, started by simply sitting in the pitch black for forty minutes, allowing his visual system to adapt to the darkness and reach maximum sensitivity. Afterward, he leaned back on a headrest and clamped his jaw down on a bite bar to lock his face into position, so photons would enter his eye at an angle and target the rod-rich parafovea. Then the volunteer pressed a button, firing the first photon, and the experiment was underway. Weeks passed, months passed, more than 30,000 trials exposing the human eye to a single photon passed. The team was ready to collate the data.

Vaziri's experiments suggest that a lone photon of light might be not simply detected by the human eye but also processed by the brain and consciously experienced. However, not every photon leads to perception. The result shows that the probability at which a human can register a photon is just above chance. Vaziri explained,

> We found that in an alternative forced choice setting, humans perceive single photons 51.6 percent of the time. The importance is not *how much* above 50 percent the results are, but that they *are* above 50 percent with statistical significance and statistical power, making it highly unlikely that the result is a consequence of random measurement fluctuations.

He added, "From the point when a photon hits the eye, there are many possible outcomes." It could be absorbed by the cornea, scattered within the eyeball's vitreous away from the retina, arrive at the parafovea but still miss a rod, hit a rod but not be absorbed by the rhodopsin molecule. "But we now know that a rod photoreceptor cell *can* respond to a single photon and even resolve the statistics of photon numbers in weak flashes of light; that this *can* trigger a biochemical cascade through the post-rod circuitry and, ultimately, this *can* lead to our perception of a single photon."

When the volunteers were asked to describe such perceptions, they agreed unanimously that while it was difficult to put into words, their experience was far from a flash of light. One said, "If you've ever looked at a dim star in the night sky and one second you see it, but the next second you don't . . . it's kind of like that." From another: "It's more like a feeling of seeing something, rather than really seeing it." Vaziri added, "It's even more extreme than that. It is a feeling at the very threshold of your imagination—a feeling that there could have been something but you aren't entirely sure." These words interrogate the gray area between the detection of light and the experience of sight. Yet what may have felt like a leap of faith is based in fact; at times, the volunteers *were* able to identify the faintest light possible. "People have been thinking for a long time in terms of an absolute threshold to human vision," Vaziri told me. "They assumed there is a hard limit in photon numbers below which light cannot be perceived. It turns out, after all, that there is no threshold. We can, on occasion, even perceive single photons." Andrew Stockman remains unconvinced by the results. "I would like to see the experiment replicated," he told me, "and I think that the event is detection rather than perception, but it is definitely interesting." Little by little, we are finding the words to reassure children who fear encroaching shadows and nighttime. It seems we have astonishing potential for vision in the darkest of places. There are moments when we even detect, albeit vaguely, the individual elementary particles that make up our universe.

✳

Back onboard FS *Sonne*, with the strange spookfish preserved in various aldehydes and safely stored, the exploration had continued. A full moon lit the night sky; its reflection shimmered on the still sea. The researchers were out on deck making the most of the remaining few days. They were heaving in their latest deep trawl, turning off the lights on the boat to protect the sensitive eyes rising from the deep. Working without head torches, they relied on their dark vision. When the moon slipped behind a cloud, leaving only stars, they could still see, if more faintly.

Our two types of photoreceptors give us surprising flexibility; whereas our cones bring us bright, cloudless days arced with rainbows, our rods give us starless and forest-canopied nights. "We are vision generalists," said Ron Douglas. "Unlike the spookfish, we see within an enormous range of lighting conditions, but it sees in conditions that leave us struggling." Vaziri's experiments suggest exceptional sensitivity, but in the photon-lean deep, vision—no matter how poor, how incoherent—still eludes us. We may share rhodopsin-loaded rods with the spookfish, but large pupils, tapeta, mirror-rigged pseudo-eyes, and four rod-packed retinas make its eyes exceptional light traps. They maximize the chances of photon hits and phototransduction events to let there be light and, ultimately, some form of sight. This leaves us to wonder how this resolves within the skull of a "four-eyed" spookfish. When I asked Douglas what it might be like to see through their eyes, he smiled and said, "I'll leave that one for the philosophers." Such questions—first posed by the American philosopher Thomas Nagel in his celebrated 1974 essay, "What Is It Like to Be a Bat?"—demand that fact make way for fancy. The mantis shrimp has exposed the difficulties of seeing through another creature's eyes; to do so through those of a "four-eyed" creature, dredged up from liquid deep space, stretches even our imagination to its limits.

A call rang out across the *Sonne* deck. Someone had spotted the lander breaking the surface, having returned from a brief stay on the ocean bed. The scientists hooked the contraption and retrieved its footage from a depth of 10,000 meters (6 miles). They stood around a monitor, craning to watch the first images from this previously unexplored frontier. The deep is slowly giving up its secrets. In Douglas's words, "Each amazing picture that comes back from the abyss is just a tiny snapshot of what is going on down there, but we're still in the dark about what lies beyond the light."

3

The Great Gray Owl and
Our Sense of Hearing

To the ancient Greeks, the owl symbolized wisdom, but the Romans saw it as an evil omen. Their myths tell of an owl-like *strix* that stalked the night and preyed on human flesh. Ovid's poem *Fasti* describes how such a demon slipped into the nursery of the sleeping prince Proca and was found hunched over the cradle, sucking the newborn's blood. This supernatural owl changed over time. In Italian, *strix* became *strega*, meaning witch; in Romanian, *strigoi* is a vampire; and, in *Macbeth*, Shakespeare once more recast the owl as "the fatal bellman" whose shriek summons King Duncan's death. Like its legendary counterparts, the great gray owl, *Strix nebulosa*, inhabits the shadows. It lives in the icy north, in the dense, dark conifer forests of Russia, Alaska, and Canada. By night, it hunts. Scythe-like talons and hooked, knife-sharp beak make it a fearsome predator. By day, it stays hidden. Although one of the largest of its kind, its dusky and mottled plumage blends with the tree branches to atomize the bird's silhouette, making it as nebulous and insubstantial as mist. Moreover, on a still moonlit night where snow blankets the landscape and deadens sound, the owl swoops on its quarry and barely breaks the silence.

The quietness of the owl's flight is unrivaled; its wing beat makes a sound so soft that it is nearly imperceptible. "While we've known this for centuries," said Professor Nigel Peake of the University of Cambridge, "what hasn't been known is *how* owls are able to fly in silence." His laboratory is one of a few around the world trying to learn from this avian acoustic stealth. For years, the focus had been the feathers along the wing's leading and trailing edges. Those at the front have tiny stiff barbs that point forward like the teeth of a comb, whereas those at the back are flexible and fringed. They work together to break up, then smooth the air currents as they flow over and off the wing, damping down any noisy turbulence. Recently Peake homed in on a third element: the wing's luxuriant touch. "We were among the first to think about the aerodynamics of this velvet," he told me. In 2016, he collaborated with scientists in America for a closer look at the smooth surface of wings from various owl species, including the great gray. They saw that the birds' primary feathers were covered with a millimeter of fine fluff. "Microscope photographs of the down show it consists of hairs that form a structure similar to a forest," he explained. "The hairs initially rise almost perpendicular to the surface, but then bend over in the flow direction to form a canopy." This Lilliputian "forest" reduces pressure fluctuations and turbulence dramatically as the air flows over the wing. The researchers, funded by the US Office of Naval Research, re-created this topography in plastic. Testing their prototype in a wind tunnel, they found it reduced sound so well that they patented the design. This discovery promises not simply stealthier surveillance aircraft or submarines but also a significant drop in everyday noise pollution from, say, wind turbines, computer fans, even the passenger planes daily crisscrossing the planet. "Owls have much to teach us about making our own world quieter," said Peake. "No other birds have wings that scatter sound so their prey can't hear them coming." The great gray is neither seen nor heard, and this natural specter also seems endowed with a supernatural sense. From a distance of some 30 meters (100 feet), it can pinpoint mice or voles with uncanny precision, even those hidden beneath mounds of virgin snow.

✳

In 1963, Masakazu, or Mark, Konishi attended a lecture at a conference on animal behavior that would end up dictating his research for the next five decades. Scientists had suggested various senses at work when owls track prey at night. Some had proposed smell, others the detection of body heat. Vision researchers saw the rod-dense retinas of the owl's enormous eyes and stressed the importance of night sight. The lecturer that day, biologist Roger Payne, was convinced that owls rely on their ears. His evidence was the spectacular snow plunge of the great gray owl. He argued that the thick drifts that conceal their prey would erase all sensory signals except sound. "Roger was such a great speaker that I was really impressed," recalled Konishi. So it was that three years later, on taking up a post at Princeton University, he put out a call for winged subjects: practicality would dictate a species of owl more readily available in New Jersey than the great gray.

A week or so later, a local bird-watcher knocked on the laboratory door, holding a box under his arm. Inside were three barn owls, only days old. Konishi learned that the chicks had been rescued from an abandoned church. They were ravenous. Colleagues helped source suitable sustenance, and before long, he was busy playing the role of owl parent. "I fed them chopped-up mice, sometimes all night. They grew. So as soon as the owls could fly, I moved them to a large room in an old house on campus." With the owls safely in their new home, tucked into a nest box, he set about finding student volunteers to share the time-consuming role of hand rearing. Soon the birds were so tame that they flew when summoned. A year later, a student knocked on Konishi's office with some unexpected news. "He told me that one of the owls had eggs. I was really surprised that owls would breed so easily in captivity." In only a few years, three would become twenty. It was time to test Roger Payne's theories on owl senses.

The studies started in a pitch-black room with a camera, infrared strobe lights, an owl also named Roger, and a tethered mouse. "I didn't want him to catch free-moving mice because mice escape and I'd be in trouble," he explained, but Roger was a quick learner and in no time was diving on the mice in the dark with lethal efficiency. However, Konishi's first images looked less promising. "It appeared as if the owl was looking at the tethered mouse as it was about to strike. This worried me a little." To establish which sense Roger was using, Konishi changed tack with a version of one of Payne's early tests. He laid foam rubber on the floor and tied a piece of paper to the mouse's tail. Now, as it scurried about, the yielding floor absorbed the pitter-patter of feet, but the paper rustled gently in its wake. This time when Roger took aim, he missed the mouse but hit the paper. In one swoop, the experiment revealed how the owl was sensing its object. "Besides demonstrating that the owl cannot see the mouse, it proved that the owl cannot locate the mouse either by its smell or by its body heat," said Konishi. "I proved Roger Payne was right in concluding that this species of owl could catch prey in total darkness using sound." Owls have ears so keen that they detect the faintest rustlings from under heavy snow, many meters away. But what are these sounds made of, and what is it exactly that the owls are hearing?

✳

"There are plenty of blind animals," observed the neuroscientist Seth Horowitz. They tend to be in lightless places—think of the blind lobsters of the deep ocean, the blind river dolphins from the muddy waters of the Indus, or the eyeless salamanders, shrimps, and spiders of caves. "Plenty of animals have very limited senses of smell," he added. "Animals such as the armadillo have a limited sensitivity to touch; and we can but hope that vultures have a limited sense of taste. But here's one thing you never find: deaf animals. Why?" In his book of the same title, Horowitz described hearing as "the universal sense." Sound is anywhere there is energy and matter, which on our planet means everywhere other than in a vacuum. Sound is simply vibration. With a twitch of its nose, a mouse inadvertently sends air molecules rippling out in all directions. Forward motion pushes these molecules closer together, creating an area of high pressure out front, leaving an area of low pressure behind. These pressure waves undulate through the atmosphere and on reaching an eardrum—whether that of another mouse, a person, or an owl—they set in motion a series of events that potentially leads to hearing.

Two qualities of sound waves are heard. The distance between the wave peaks creates the pitch or note, measured in hertz. Waves in quick succession, high frequency, are perceived as high notes, whereas waves coming in more slowly, with peaks spread out and lower in frequency, are perceived as low notes. Meanwhile, the pressure of the oncoming wave influences the note's volume, measured in decibels. The ability to hear may be universal, but what is heard is not: creatures have different ranges circumscribed by pitch and volume. The lowest note that we humans perceive is a rumbling 20 hertz, the highest is a piercing 20 kilohertz, and we hear best between 1,000 and 4,000 hertz. We are deaf to the 17-hertz low thunder of elephants communicating across long distances and to the 120-kilohertz high screech of bats swirling above our heads; these infra and ultra sounds are below and above our audible register, respectively. When it comes to our range of volume, our pain threshold dictates the upper limit.

On August 27, 1883, when a volcano tore apart the island of Krakatoa in Indonesia, it unleashed the loudest sound in recorded history. Although its pressure, and therefore volume, decreased over distance, the sound could still be heard 4,800 kilometers (3,000 miles) away. The local police chief of Rodrigues, in the Indian Ocean, reported it "coming from the eastward, like the distant roars of heavy guns." Imagine a noise originating in Dublin, Ireland, being heard across the Atlantic in Boston, Massachusetts. Yet well before that, at only 65 kilometers (40 miles), it reached the British ship *Norham Castle*, where it surpassed the human hearing threshold. The captain's logbook recorded, "So violent are the explosions that the ear-drums of over half my crew have been shattered. My last thoughts are with my dear wife." The loudest pressure wave our eardrum can register before rupture is around 130 decibels, but any above 85 can cause hearing loss. Conversations take place at around 60 decibels. Rustling leaves or whispers hover at 20 decibels. Yet it is possible to hear even more muffled sounds, as evidenced when stepping through the sealed doors of the quietest place on earth.

*

In 1951, the composer John Cage traveled to Harvard University in search of silence. He had been given permission to spend time in Beranek's Box. This structure, built during World War II by the renowned acoustics scientist Leo Beranek, was probably the world's first echo-free, or anechoic, chamber. He had filled a concrete cube with so many fiberglass wedges that he shrank its three dimensions from 15 meters (50 feet) to 2 meters (6.56 feet). The fiberglass absorbed noises from within; thick walls insulated against those from without. Cage entered the room and stood on the metal grille suspended between the sound-deadening surfaces. The door closed, and he was left alone. As his ears adjusted, he waited for a hush to descend, the negation of sound, but it did not come. He later recounted, "In that silent room, I heard two sounds. One high and one low. Afterwards I asked the engineer in charge, why, if the room was so silent, I had heard two sounds." Cage was told that the high one was his nervous system and the low one was his blood circulating.

Over the years, others have spoken of similar experiences in rooms designed since that are even quieter than Beranek's Box. Gentle inhalations and exhalations, around 10 decibels, are said to sound like Darth Vader's stertorous rasps; heartbeats, mere decibels, thud. As minutes pass, these anechoic explorers then describe sounds beyond these more obvious bodily commotions. "After about twenty minutes, I began to hear a high-pitched whine, which persisted," claimed one, "probably the sound of my circulatory system." Another, "I started to hear the blood rushing in my veins. Your ears become more sensitive as a place gets quieter, and mine were going overtime. I frowned and heard my scalp moving over my skull, which was eerie, and a strange, metallic scraping noise I couldn't explain." Whatever the source of these sounds, as Cage said, there is always something to hear: "In fact, try as we may to make a silence, we cannot." He added, "There is no such thing as silence." This revelation inspired his perhaps most famous composition, 4″33′: four minutes, thirty-three seconds of a pianist not playing, to draw the audience's attention to the lack of silence in the world.

Ultimately, Beranek's Box and the more muted anechoic chambers built since emphasize the extraordinary sensitivity of the human ear. We can hear zero decibels. Whereas 60-decibel conversations displace particles in the air by 10 millimeters and 20-decibel whispers by 0.0001 millimeters, a zero-decibel pressure wave—0.000002 pascals to be precise—vibrates our eardrum by a mere 0.00000001 millimeter. As the sensory researcher Pete Jones said, "To put that in context, 0.000002 Pascals is less than a billionth of the ambient pressure in the air around us, and 0.00000001 mm is smaller than the diameter of a hydrogen atom!" Our ability to hear is superb. "We wouldn't want our auditory system to be much more sensitive, since then we would constantly hear the hum of atoms vibrating in the air." So how does the creature with a flair for detecting mouse-sized vibrations compare?

✳

The aim of Mark Konishi's next study was to uncover the limits of owl hearing. It would require making his lightless experimental chamber anechoic, along with three particularly pliant owls. "Since owls respond better at night, day and night were reversed," he said. "A bright white light simulated daytime between 9:00 p.m. and 9:00 a.m." It was then replaced by a dim red light to fool the owls into thinking it was time to stir and stretch their wings. "The first step in the training procedure was to teach the owl to obtain food from a semiautomatic feeder." This wooden box, with a beak-sized slot in its lid, contained a circular plexi-glass tray bearing cups brimming with mouse morsels. To reward the owl, Konishi needed only to switch a motor that moved the tray so a cup slid into alignment with the opening. He trained the birds to leave the perch when they heard pure tones of a certain frequency played through a loudspeaker. "It usually took ten to fourteen days before the owls learned this task reliably." Konishi played different frequencies of between 500 and 15,000 hertz, methodically reducing the volume each time until the owl failed to respond. The owls were tested in turn; the work took months.

At last, Konishi could map a precise hearing curve, an audiogram that graphically represented the most hushed sounds an owl can hear across the full frequency spectrum. He discovered that barn owls hear lower volumes, across a greater range of frequencies, than almost any other bird, fish, amphibian, or reptile. He also learned that although mammals hear much the same range, owls hear nearly all the notes with greater sensitivity, apart from the few above 12,000 hertz. "I compared people and the owls under the same conditions," explained Konishi. "The owls could hear sounds that were so faint that none of my young undergraduate students and assistants could register them at a distance of the owl's perch." Owls could hear 20 decibels below our limit. Konishi began to question how such range and unparalleled sensitivity were possible.

*

Two chance meetings some thirty years ago would change Christine Köppl's life. "My boyfriend at the time was Mark Konishi's first PhD student, and he took me to Mark's lab where I first encountered the owl. A lot of work had been done on the owl's incredible sense of hearing and I realized few had looked at the hearing organ itself." The boyfriend would become her husband, and Köppl—now a professor of neurobiology at the University of Oldenburg in Germany—became a rare authority on the owl's ear. Owls do not have visible external ears like ours. Instead, their ear openings are hidden, discoverable only by parting the feathers behind the ruff that surrounds their face; those of the great gray are so large that we can peer through and even see the backs of their orbital bones. "The other remarkable aspect of these and other owl species is that their ear openings are not at the same height," Köppl told me, "so looking face on at the barn owl, for example, their right ear is below eye level at about 8 o'clock, whereas their left is at 2 o'clock." That said, once gathered by an external ear or not, a sound wave's journey is markedly similar in owls and humans. It funnels down an auditory canal, then hits and vibrates an eardrum, thereby moving tiny bones within the middle ear called ossicles. These drum on the coiled bony labyrinth of the inner ear, named the cochlea, from the Greek word *kokhlías*, or snail. "The owl cochlea, like ours, is the business end of the ear," Köppl said, "and it has fascinated me since I first laid eyes on it."

Köppl would go on to pioneer the use of electron microscopy to unveil the owl cochlea in unprecedented detail, but even before she looked closely, there were two features that struck her as odd. The first was its outline. "Most birds have a banana-shaped cochlea," she explained, "but the owl's was a more complex three-dimensional shape. It actually looked warped." She saw how it curved around the bird's braincase. "It is widely believed that the coiling snail shape of our cochlea evolved to fit its length into a small space, so I wondered whether the reason for the owl's strange shape was its unusual size." This was its second peculiarity. "It was enormous," she told me. "Our measurements showed an average of 12 millimeters [0.5 inch]. That's very, very large. It's by far the longest for any bird studied and by some margin." Köppl decided to look inside the bony casing at the sensory tissue.

All vertebrate cochleae are filled with a fluid called endolymph, through which threads a basilar membrane covered in microscopic hair cells. When the ossicle hammers on their outside, waves are generated inside, within the endolymph. These wash along the length of the membrane, gently bending and exciting its hairs. Köppl said, "These hair cells are where sound vibration is transformed into an electrical signal to the brain and begins the perception of hearing." They are sound's sensory cells: hearing's equivalent to vision's cones and rods. They are where the outer physical reality starts its transformation into inner experience. Köppl and her colleagues extracted a collection of owl basilar membranes and placed them beneath the microscope. With great care, they counted every sensory hair cell from base to tip. She recalled, "The average tally for the barn owl was 16,000 hair cells, thousands more than in any other bird." Next, electrophysiological studies showed how sound frequencies are sorted along the membrane. The researchers saw that high frequencies excited its base, leaving low frequencies for its tip—nothing unusual—but one narrow band of sound was allotted such a disproportionate stretch that it accounted entirely for the extra length Köppl had seen. "More than half of the cochlea was devoted to processing frequencies between 5,000 and 10,000 hertz," she explained. "We realized we were looking at an auditory fovea. This surprised me because it had only been seen once before, in bats. It had never been seen before in birds and has not been seen since." Just as the cone-dense visual fovea of our retina heightens sensitivity in one area of our field of view, this cochlear adaptation makes bats and owls receptive to certain sounds. The bat fovea attunes them to the reflections of their high-pitch ultrasonic calls, facilitating echolocation, whereas the owl fovea attunes them to an octave higher than other birds, and to the rustlings of their prey. Although the discovery explained the owls' extended hearing range, the question of their exceptional sensitivity remained unresolved.

Surveying the owl hair cells under vast magnification, Köppl was reminded of those in another cochlea: "The anatomical ultrastructure of the owl's hair cells and how they were dispersed along the basilar membrane bore many resemblances to our ear." Her observations fed into an international research effort that is disclosing astounding similarities between the inner ears of owls and of humans:

> They both have two specialized groups of hair cells. Humans have inner and outer ones; owls have tall and short. Human inner and avian tall hair cells transmit to the brain and create the perception of sound, whereas human outer and avian short hair cells likely amplify the signal and make the system more sensitive. Overall the resemblances are so strong they suggest that hearing underwent an independent but parallel evolution in birds and mammals.

The commonalities also meant that Köppl found little in the owl's inner ear to account for its auditory edge. "The basic hair cell cannot be more sensitive," she explained. Owls may have more hair cells than other birds, but they would need vastly more to account for hearing 20 decibels below us. "They do not have the numbers." All of these observations led Köppl to conclude that "the inner ear of the owl is as sensitive as those of humans." Their eardrums do not register smaller disturbances. Their ears are not overwhelmed by the hum of atoms vibrating. An owl's keen hearing has less to do with the sensory part of its ear than—like the bird's silent swoop—its plumage.

"We now understand that the owl's acute sensitivity is thanks to its pretty face," Köppl told me. Densely packed feathers form its concave surface and give the barn owl its distinctive heart-shaped face. "When one sees the whole design of the facial disc, one cannot help thinking of a sound-collecting device," said Konishi. "And there is compelling evidence to support this," Köppl added. "Various experiments have shown that when the ruff is removed, the barn owl loses its 20-decibel advantage and its hearing threshold matches ours." The great gray's disc is as well developed and even bigger than that of a barn owl. Although the experiments have not been done, Köppl told me that "this will give the great gray owl an auditory sensitivity at least as great as the barn owl." Much as the mirror extends a spookfish's field of vision, the wide feathered ruff extends an owl's field of hearing. It captures sounds from the surroundings, then reflects and channels them toward an owl's ear openings. "Essentially one can think of an owl's facial disc as analogous to our outer ear, but because of its size, it acts more like a parabolic reflector collecting more sound." Given that our inner ears are comparably sensitive, holding up a Victorian hearing trumpet to, or cupping a hand around, our outer ear will increase our sound scope. These simple actions pluck soft sounds from far away to ensure they rattle our eardrum and excite our inner ear, affording us some insight into the owl's whispering world. If the symmetry in the sense of hearing between humans and owls is striking, the Oldenburg team would soon discover a significant imperfection in the ear of their oldest bird.

"In 1993, Christine gave us some nestlings, which enabled us to start the Oldenburg Barn Owl Project," zoologist Ulrike Langemann told me. Thus began one of the longest-running scientific studies into barn owl hearing. "Weiss was only twelve days old at the time and we named her for her lovely white face." As Konishi had done before, the team hand-raised the birds. "We trained them like falcons to fly to us when we showed them the glove, the meaty treat, and called for them." They tied leather jesses to the owls' ankles to hold them securely when being walked between the aviaries and experimental chambers. "Weiss was shy like most barn owls but so easy to work. She was a favorite," admitted Langemann. "She even remembered me after two study years abroad. Barn owls are like elephants; they never forget, especially if you're the one who feeds them."

The project started by replicating Konishi's threshold experiments. Weiss had her first hearing test a few months before her second birthday—as did all the other owls—and these continued down the years. Then in 2017, Langemann realized that she was sitting on a scientific gold mine. "These threshold data going back over two decades were just gathering dust in the drawer. Put together, it could tell us how an owl's hearing ages." Langemann had data on three groups of owl siblings, including Weiss, so seven birds in all. She compared the tests that had been done before and after they reached ten years. "Weiss was twenty-three years old the last time I tested her." In the wild, owls rarely live beyond four. "Her eyesight wasn't good, and I was expecting deterioration in her hearing too, but not at all." The other birds seemed to follow Weiss's example. "The old owls' hearing was as good as the young's. Frankly, their lack of hearing loss was remarkable." Scientists with interests typically beyond owls are intrigued by Weiss's ageless ears.

Our, on average, 20,000 cochlear hair cells are paltry in number when contrasted to the retina's 100-million-odd rods and 5-million-odd cones. Inevitably, all sensors accrue damage over the years, but whereas reducing our photoreceptor millions by a thousand makes no odds to vision, the same reduction to hair cells causes irretrievable hearing loss. Arguably, hearing is the most fragile of all our senses. In humans, deafness is the scourge of the elderly. Presbycusis, or age-related hearing loss, is so common that it affects one in three people over sixty-five and one in two over seventy-five. "There is plenty of exciting research coming out to show why old birds like Weiss have pristine hearing thresholds and it is because of their hair cells," Köppl explained. Research is underway to understand bird hair cells and the secret to their youthfulness. Scientists have found that they never stop growing and mending. "There is a lot of speculation as to how and why birds can regenerate, but this fact, alongside our knowledge that these hair cells are so similar to ours, makes research into this area additionally attractive," Köppl said. "I believe it may well one day help us cure deafness." Langemann added, "Weiss passed on soon after we published the paper. I still miss her, but what she showed will outlast us all." Only time will tell how the owl and her extended feathered family will help those who don't hear well. Meanwhile, she already sheds light on the experience of being blind.

*

John Hull's tribulations began when he was a young boy in Australia. At the age of thirteen, he developed cataracts in both eyes. Operations restored his sight but weakened his retinas so much that gradually, over the years, they detached. Eventually, as he described in his memoir, *Touching the Rock*, this left him "deep blind": not only without vision but also without memories of vision or light. Nevertheless, as his sight waned, he began to notice the emergence of another unexpected sense. It started with the loss in his left eye: "It took the form of a sudden, vivid awareness of an object on my blind side, within a few inches of my head. Stepping out to cross the road, I would recoil from something immediately on my left. Glancing round, there would be something like a parked van." This feeling returned more palpably in his forties with the failure of his remaining eye. Having moved to Britain, where he received a PhD in theology, Hull was working as a lecturer of religious education at the University of Birmingham. At the end of each day, he would wait for the campus to empty before navigating his way home. "In the quiet of the evening, I had a sense of presence, which was the realization of an obstacle. I discovered that if I stopped when I had this sense and waved my white cane around, I would make contact with a tree trunk." Over the ensuing weeks, the sensations recurred. Bulky shapes like people or parked cars were more obvious than slender lampposts, and the awareness did not extend beyond five feet. "I gradually realized I was developing some strange kind of perception," he recounted. "The experience itself is quite extraordinary, and I cannot compare it with anything else I have ever known. It is like a sense of physical pressure . . . upon the skin of the face."

Hull was far from alone. This enhanced perception has long been a common experience of blind people. In 1749, the French philosopher Denis Diderot attributed this "amazing ability" to the action of air on the face. In 1905, another Frenchman, the ophthalmologist Emile Jarval, decided it was a sense other than sight, hearing, touch, taste, or smell and named it by coining the term *sixth sense*. It has since gone by many names: telesthesia, paraoptic vision, and mostly, because of where it is felt, facial vision. However, pioneering research from the 1940s showed it has nothing to do with the face.

A silent, monochrome movie exists of Karl Dallenbach's experiments in a vaulted, wood-beamed laboratory at Cornell University. A man in dark glasses with a white cane is offered a cigarette while presumably being talked through the plan. Dallenbach blindfolds two sighted and two blind individuals. The film continues with the subjects being guided in disorienting circles before being left to walk alone down the long room toward a wall. The rolling titles read, "The end wall was used because it gave a large reflecting surface. Subject raises the left arm when the wall is first perceived; the right arm when the wall is judged to be only a few inches away." Although the sighted people appear to step out with more hesitation, like their blind colleagues, they stop short of the wall. Dallenbach suspected they must be using the same cues as one another, but that those cues had nothing to do with the pressure or temperature of air on skin. Having noticed the noise their shoes made on the hardwood floors, he speculated a sense other than touch at work. He asked the volunteers to repeat the procedure without footwear, initially on a rolled-out carpet and then wearing headphones. The footage shows a blind man, wearing both blindfold and headphones, walking down the carpet runner. This time he approaches the Masonite end board more tentatively but nevertheless collides with it. Another title states, "All the subjects failed to detect the obstacle in every one of the 400 trials." Dallenbach's final experimental flourish involves a man walking the course once more but with a microphone held before him. The blind subject, still in headphones, sits sequestered in another room; he listens to the sounds gathered by the walker's microphone. Through sound alone, he detects the approaching wall from feet away and displays such accuracy that he then allows the man to walk within an inch of it. Dallenbach had proved that facial vision relies on the ears.

*

There is a facet to hearing that is often overlooked. This sense does not simply gauge the qualities of sounds—their tone, timbre, or volume—

but also delineates surrounding space in unexpected ways. This quality becomes more apparent when sight fails. Most people might think a pleasant day is one when the sun shines, but to John Hull, it would have entailed a light breeze. "This brings into life all the sounds in my environment. The leaves are rustling, bits of paper are blowing along the pavement." Or perhaps a storm: "Thunder puts a roof over my head, a very high vaulted ceiling of rumbling sound. I realize that I am in a big place, whereas before there was nothing at all." And to him, a cloudburst was like light falling on a landscape:

> Rain has a way of bringing out the contours of everything; it throws a colored blanket over previously invisible things; instead of an intermittent and thus fragmented world, the steadily falling rain creates continuity of acoustic space. . . . The world, which is veiled until I touch it, has suddenly disclosed itself to me.

As Karl Dallenbach showed, should darkness engulf any one of us, we would remain able to envision aspects of the world through sound. His experiments have been repeated successfully and adapted through the years. Scientists have shown we hear silent objects through the sounds they reflect; we can even hear their general shape and the material from which they are made. Although blind people are generally better at such bat-like echolocation, anyone can learn this skill with relative speed, and some scientists suggest we already unconsciously use a degree of echolocation in our day-to-day life. We are also experts at pinpointing where sounds come from. As a matter of fact, our species held the record for the greatest known ability to locate sounds, prior to Mark Konishi testing the owl. To hunt by sound alone, these birds home in on their prey with such precision that their hearing has been described as earsight.

Konishi transported twenty-one birds across America to take up a new post at the California Institute of Technology. The Caltech machinist Herb Adams—who had worked on the Viking lander for NASA's first mission to Mars—was designing a revolutionary acoustic room to help him investigate the owl's talent not simply for hearing soft squeaks but also for processing and locating their origin. "In the end Herb built a bowl-shaped hoop and installed a system to move a small loudspeaker along the hoop so that we could place the speaker anywhere around the owl's head," said Konishi. "Herb's hoop," as it became known, enabled Konishi's team to interrogate how an owl sits on a perch in a soundproof room while its brain decodes sounds arriving in three dimensions. "Owls determine the location of a mouse so quickly, in the first two hundred microseconds, a fifth of a second, so they are not spending a lot of time doing auditory computation," he explained. "So I dreamed of finding a map of auditory space in the brain: a map that would turn sound into space." Konishi did not conceive of the difficulties ahead, as nothing like this had been found before. He admitted, "This is where ignorance or naiveté helped." With the owls settled in their new home, the research got underway. "From the first experiment, we had some encouraging news; we found neurons in the brain with very sharp directionality." Sounds from different places were activating different areas of the owl's midbrain. "Then several weeks later, we could see a pattern of change as we recorded from different locations." Despite the quick start, the ensuing work would take several months, then years; ultimately, Konishi and the team were able to expose the science behind the owl's earsight. They uncovered the first sensory map in the brain that represented information in a way an eye could understand—"a retina-like map"—but relied on the bird's two ears.

In *Musicophilia*, Oliver Sacks recounted the story of his friend Howard Brandston, who, after a sudden attack of vertigo, suffered a near-complete loss of hearing in his right ear. "I could still hear sounds on that side," he told Sacks, "but could not unscramble words or distinguish tonal differences." His life was affected in unexpected ways. "The following week I had concert tickets but the musical performance sounded flat, lifeless and without the harmonious quality that I loved. Yes, I could still recognize the music, but instead of the uplifting emotional experience I was anticipating, I became so depressed that tears came to my eyes." Soon another unforeseen complication arose. Brandston, a keen outdoorsman, had set off on his first deer-hunting expedition since his hearing loss. "Standing absolutely still, I could hear the scurry of the chipmunk, the foraging squirrel, but the ability I formerly had to pinpoint the location of these sounds was now lost to me." It was only then, when he could not hear the whereabouts of his quarry, that he realized how much the hunter is dependent on two working ears. Sacks drew parallels to those who have become blind in one eye and lose the ability to see depth stereoscopically. He suggested, "The resonances of losing stereoscopy can be unexpectedly far reaching, causing not only a problem in judging depth and distance, but a 'flattening' of the whole visual world." When listening to music, Brandston's soundscape had become similarly flattened, as if it had lost its architecture. When hunting, he could see the landscape, but he could not sense the position of the sounds within it.

"Our ear and the owl's ear break down complex signals in a similar way," said Konishi. "Quite how had been guessed at in humans years ago, but the owls proved how humans locate sound." He turned once again to Herb's hoop, but this time plugged each of the owls' ears in turn. The birds' responses slowly disclosed that there are two processing pathways. "From the inner ear, the auditory nerve projects onto the brain and each fiber in the auditory nerve divides into two branches." The team found that these two branches exploit subtle differences in a sound reaching the left and right ears. The first compares the difference in arrival times. This is greatest when a sound comes from either side; only a noise from directly in front arrives at both ears at the same time. "The owl uses microsecond time differences," Konishi explained. "The maximum time differences these barn owls detect are only 150 microseconds, and the smallest differences that these owls detect are 10 microseconds": that's 150 millionths and 10 millionths of a second, respectively. As with all ground hunters, the owl must be able to locate its prey on the flat plane, so the first pathway uses these microsecond differences to compute a sound's horizontal location. Since the owl approaches from the air, it must also be able to determine its angle of elevation above the target. The second pathway compares tiny differences in volume, making use of the owl's asymmetric ears. If a sound originates above the horizontal line of sight, it will be louder in the lower right ear than in the higher left. Only a noise at eye level has the same volume. The scientists had found a place in the bird's brain where the time and volume pathways converge: where their horizontal and vertical data merge into a three-dimensional map of sound space.

Konishi demonstrated that when it comes to locating sounds in the landscape, the owl's earsight is unmatched. Of all the owls, as the ornithologist Tim Birkhead said, the one "that best captures the extreme sophistication of avian hearing is the great gray." As this feathered phantom flies over an unmarked swath of snow, its facial parabolic antenna picks up the distant scurryings of a mouse. The deep silence of its flight ensures these sounds are not masked. The owl pinpoints their source in three dimensions from the small differences in how they arrive at each of its asymmetric ears. The owl continually listens and adjusts its flight path. Without the need of sight, smell, or any sense other than sound, it will zero in on its victim at speed. As humans, we exploit the very same subtle sound cues, determining location from the time and volume differences between our ears. With our feet firmly on the ground, we rarely need to compute the angle of elevation—indeed, on this axis, we are three times worse than an owl—but in the horizontal world that we inhabit, we are just as accurate. Like the bat, we hear our world through the sounds it makes and reflects; like the owl, we map these to hear space.

*

Scientific research has coaxed the owl from the shadows and restored her to Athene, the Greek goddess of wisdom. Through this creature, we learn what it means to hear: not simply to detect sounds but to create rich and perspectival soundscapes. We discover our talent for discerning whispers of whispers, then locating and layering them to build cathedrals of sound. The silent bird also guides us toward making this world a better place: whether through redesigning technology to subdue unwanted noise or improving the lives of those less fortunate. "I am just as deaf as I am blind," wrote the American deaf-blind activist Helen Keller to her doctor in 1910. "The problems of deafness are deep and more complex, if not more important, than those of blindness. Deafness is a worse misfortune." The owl sits on the shoulder of the blind, bearing the gift of earsight. One day, alongside her wider avian family, she may also offer others the gift of sound.

4

The Star-Nosed Mole and
Our Sense of Touch

T HE ACCOLADE FOR FASTEST KILLER IN THE ANIMAL KING-
dom goes not to the cheetah but to an inhabitant of the wetlands of the northeastern United States and eastern Canada.
Guinness World Records hails it as the most rapid and voracious mammalian predator. The entry reads, "It takes humans 650 milliseconds
to respond to a red light when driving." An eyeblink lasts at most 300
milliseconds, yet this creature can identify, capture, and consume its
prey in less than half this time. One wonders whether its unfortunate
quarry even has time to register its fate. To compound the creature's
sinister reputation, it has been compared to H. P. Lovecraft's famous
extraterrestrial Cthulhu: "A monster of vaguely anthropoid outline,
but with an octopus-like head whose face was a mass of feelers, a scaly,
rubbery-looking body, prodigious claws on hind and fore feet, and
long, narrow wings behind." Aside from the scales, rubber, and wings,
the features of the subject of this chapter appear eerily similar: powerful shoulders with enlarged and clawed forelimbs and—in one of the
most startling adaptations in the animal kingdom—twenty-two blood-
engorged, fleshy tentacles that radiate from its nostrils. Yet this furry
fiend barely fills a human hand, seldom ventures aboveground, and
poses little threat unless you are an earthworm. It is a rather bewitching
mammal, whose unusual snout gives it the name *Condylura cristata*, or
the star-nosed mole.

Nashville, Tennessee, is home to both country music and star-nosed
mole research. Ken Catania, the world authority on these animals, is
based at Vanderbilt University. Guided by the words of the American
theoretical physicist John Archibald Wheeler, who once said, "In any
field, find the strangest thing and then explore it," Catania met his
match in the mole. "It's hard to imagine a stranger animal," he told me,
"like a creature you might picture emerging from a flying saucer to
greet a delegation of curious earthlings." The mole's star-shaped nose
has been the subject of intense speculation since the 1800s. Some have
even wondered whether it acts like an antenna to detect electrical fields.

"As a neuroscientist interested in sensory systems, this kind of biological anomaly represents an irresistible mystery," said Catania. He collects his moles from the vast, sprawling wetlands of northwestern Pennsylvania. These half a million acres of lush vegetation and soggy soil are Elysian fields to a maker of tunnels. The secretive and nearly blind creatures' dig runs over 30 meters (100 feet) long, at a pace of 2.4 meters (8 feet) per hour, in search of tasty morsels. As the only moles known to swim, they also hunt small fish and crustacea in the swampy shallows. When Catania filmed them foraging underwater in his laboratory tanks, the footage revealed how they engulf an appetizing item with bubbles blown from their nostrils. Catania believes that when these air pockets surround the target, they absorb its scent molecules. Then, on reinhaling this odor-laden air, the moles are able to smell whether what they have found is edible. "This sensory ability was completely unanticipated," said Catania. "It was previously—and logically—thought that mammals could not use olfaction underwater, but this newly described behavior provides a mechanism." Thanks to its snout, the mole was thus the first known mammal to smell underwater. Yet this would prove the least of its sensory talents.

As a star-nosed mole advances down a tunnel, the eleven pairs of nose rays that encircle its nostrils move so fast as to be a blur. To see what was happening, Catania turned to a high-speed camera to slow them down. He built a plexiglass-bottomed burrow, baited it with well-hidden millimeter snippets of earthworms—ambrosia to a mole— and positioned his camera to catch a mole-belly view, at 1,000 frames a second. His first subject raced through the burrow, finding and devouring the pieces of worm with great efficiency. The second and third moles followed suit.

When Catania played back the footage in slow motion, he could clearly see the twenty-two nose rays on the moles' stars, feeling their way in the darkness. The rays moved quickly and independently, touching up to a dozen objects every second. Moreover, whenever one of the first ten pairs encountered a morsel, the eleventh, somewhat shrunken pair, would always investigate, before passing the prey on to the mole's tweezer-like front teeth. Catania added, "No food item is eaten without first being explored with ray eleven." The eleventh pair may be small, but its role is great. Catania and his colleague Fiona Remple also measured the time this took and found that the mole can identify and catch its prey in as little as 120 milliseconds. "A phenomenally short amount of time," claimed Catania. "I don't know of any other mammal that comes close to this." It was this revelation that led to a call from the office of Guinness World Records. "They sent me this nice certificate in the mail," Catania said. "I joke that I took down my PhD record off the wall and put this certificate up there instead." The implications of the high-speed footage are more than mere record-breaking; it shows that the mole's nose is more attuned to the sense of touch than it is to smell. This is an idea that was first proposed a century and a half ago, on another continent, with another species of mole.

Gustav Heinrich Theodor Eimer worked as a comparative anatomist at the Julius Maximilian University of Würzburg, Germany, and had become intrigued by the common or garden European mole. Eimer observed that the skin of the mole's nose was covered with a patchwork of minute swellings, each one up to a fifth of a millimeter across. "These points or papillae are the seat of peculiar nerve endings," he later wrote and, interest piqued, took a closer look beneath a microscope. His wife, Anna, an artist, recorded what he saw; her fine, ethereal drawings show that the domed papillae are columns of stacked and squashed epidermal cells with nerve bundles running down central chimneys. Eimer dyed the tissue to expose previously hidden detail. "If you treat the skin," he continued, "one discovers a surprising amount of nerves. Myelinated nerve fibers run in various directions in thick bundles." Painstakingly, he counted more than 5,000 papillae in the skin of one mole snout, with two to three nerve fibers running through each core. He came to the unexpected realization that, because "the nerve endings . . . must come into direct contact" with whatever the nose does, the papillae must enable the mole to feel its surroundings. He called them *Tastkegel*, or touch cones. In 1871, he published his findings in the scholarly periodical of the day, *Archiv für mikroskopische Anatomie*, under the delightful heading "Die Schnautze [*sic*] des Maulwurfs als Tastwerkzeug" ("The Muzzle of Moles as a Touch Organ"). With perhaps inadvertent dark humor, he noted, "This unusual amount of nerves easily explains the well-known fact that the lightest of punches on a mole's snout will kill him." He concluded that "the mole's snout must be the seat of a tremendously trained sense of touch, because it almost completely replaces the animal's sense of face and is the only guide on its underground paths." Today *Tastkegel* are called Eimer's organs in his honor. By rendering skin highly receptive to feeling, they transform the European mole's nose into an organ that privileges touch above smell.

✳

Our skin delineates us, gives us shape, and holds us together. It shields us from infections, chemicals, and the elements. It cools us down with sweat, then heats us up with goose bumps. It is also where we come into contact with the outside world, and, like the mole's nose, it harbors our sense of touch. Consequently, our skin is the largest sense organ in our body by some margin. A macabre collection at the Medical Pathology Museum in Tokyo University is a visceral visualization of this. Dr. Masaichi Fukushi, a world expert on tattoos, removed and preserved whole human hides. Splayed, stretched, and mounted, the skin of an average adult male covers some 2 square meters (6.5 square feet) and constitutes about 16 percent of total body weight. A closer look at our fingertips reveals the ridges and whorls of fingerprints. Although these are replete with sentimental significance—external marks "of human individuality written in an obscure artistic code," according to the neuroscientist David Linden—we have yet to understand their purpose.

Closer still, beneath a microscope, this skin resolves into many layers. The epidermis (from the Greek *epi* (upon) and *derma* (skin)) is made up of five. At the surface is the stratum corneum of dry, dead skin cells that slough off every few weeks; farther down is the smooth, translucent stratum lucidum, which is found only on our fingers, the palms of our hands, and the soles of our feet; then the stratum granulosum of keratinocytes, the most common skin cells; next comes the stratum spinosum, the squamous cell layer, containing the Langerhans cells that confer immunity. Finally, there is the stratum basale, with melanocytes, or pigment cells, and column-shaped basal cells that divide, pushing older cells upward. Deeper still, in the dermis, the bulk of skin is made up of a mesh of connective tissue that cocoons our nerves, blood vessels, and sweat glands. It is in these epidermal and dermal borderlands that our sense of touch begins. Although our skin shows no signs whatsoever of the mole's domed *Tastkegel*, other structures were found and described in a paper published in the same periodical as Eimer's four years later by another German anatomist.

In 1875, Friedrich Sigmund Merkel claimed to have discovered the first known *Tastzellen*, or touch cells, in mammals, including humans. Working at the University of Rostock, hundreds of miles north of Würzburg, he had identified saucer-shaped cells at the bottom of the epidermis, in the stratum basale, in close proximity to enlarged nerve terminals that ran into the dermis. Merkel believed they grant our powers of touch. The passage of time would prove him right; today, these cells carry his name. Subsequent research has demonstrated that Merkel cells react to the lightest of brushes; if the skin is depressed by a mere five-hundredths of a millimeter, they transform the physical deformation into an electrical impulse. Merkel could not know it, but his touch discs would prove to be one of four skin touch cells or mechanoreceptors. They form a supreme team of *Tastzellen* working together to convey the infinite variety of how the world feels to our brain. Each is named for the scientist behind its discovery. Egg-shaped Meissner's corpuscles, found alongside Merkel cells, detect feather-light pressure and low vibration, earning them the epithet "kissing receptors." Pacinian and Ruffini corpuscles are buried deeper in the dermis. The first, layered like an onion, reacts to high-frequency vibration; the latter detects the skin distortion and stretching that one senses when squeezing a hand into a tight leather glove. We now know that it is the Merkel cells that gauge edge, shape, and texture, enabling us to discriminate between the smooth symmetry of a ball bearing and the corrugated roughness of a walnut. Most abundant in our fingertips, they endow such nuance that we can discern features that are mere fractions of millimeters apart. In some situations, the respective discoveries of Merkel and Eimer converge.

Today Ken Catania has a tool at his disposal more powerful than his nineteenth-century peers could have imagined. The scanning electron microscope boasts magnifications at least a hundred times greater than a light microscope, enabling him to see with such topographical clarity that the mole's star-shaped nose emerged as a whole new world. "Under the scanning electron microscope, its skin surface became a cobbled landscape of tens of thousands of epidermal domes. A beautiful anatomy," he told me. Each dome represented an Eimer's organ. "I did a careful check to see if there were any other types of sensors but found nothing else." The star disclosed no evidence of electroreceptors, chemoreceptors, or any other sensory abilities beyond touch. "The Eimer's organs were packed in hexagonal arrays, like a honeycomb, so there was no room for anything else." Catania focused on a single organ, then increased the resolution. He saw that the uppermost stratum corneum was barely present at only six or seven cells thick—"only five to six microns between the external environment and the living tissues." Looking deeper, just beneath the organ's cellular column, buried on the bottom of the epidermal floor, Catania found, "in prime position to be compressed by pressure on the cell column," a cell much like Merkel had found in our skin. Seeing the mole's star crammed with Eimer's *Tastkegel* and primed with Merkel's *Tastzellen*, Catania came to the same conclusion as Eimer had of the European mole: "The star-nosed mole's 'nose' is not an olfactory organ, but a skin surface that mediates touch."

Catania set about the unenviable task of counting the domes in the mole's star and, more onerous still, each of the individual fibers within the complex nerve constellations. Each Eimer's organ was a quarter the size of those on the European mole. So it followed that there might be a few more on the star, but Catania found on average about 25,000: five times more. Given that each was served by around four myelinated nerve fibers—as opposed to the two or three core fibers on Eimer's mole—Catania reached a conservative estimate of 100,000 nerves in total: again, many thousands more. On every count, the star-nosed mole surpassed its European cousin by some margin. We humans pale further in comparison. "The nerve count on the mole's snout alone adds up to many more touch fibers than one finds on the whole human hand"—100,000 to the mole, 17,000 to us—"and yet the entire star is only a centimeter (less than 0.5 inch) across. Imagine having six times the sensitivity of your entire hand concentrated in a single fingertip." Catania concluded that there is no other mammalian organ known to science as densely innervated and acutely primed for touch as this snout. "Humans may have more diverse ways of feeling with their fingers, but the star-nosed mole has the most sensitive touch organ of any mammal yet discovered." Thanks to this celestial snout, the mole perceives the world through touch. There are rare individuals for whom this sense assumes a similar significance.

✳

Helen Keller was born at the tail end of the nineteenth century and is perhaps history's best-known deaf and blind individual. She was robbed of her sight and hearing when only nineteen months old. It is nearly impossible to imagine her daily experience: total silence, perpetual blackness, and contact with others that was more or less restricted to touch. On March 3, 1887, three months before she turned seven, she would meet someone who would widen her world. She wrote in *The Story of My Life*, "The most important day I remember in all my life is the one on which my teacher, Anne Mansfield Sullivan, came to me. I am filled with wonder when I consider the immeasurable contrasts between the two lives which it connects." Sullivan, herself partially blind, was struck by her new pupil's almost complete isolation: "She had no way of communicating with those around her except for a few imitative signs that she had made for herself. A push meant go and a pull meant come." Sullivan realized that having never heard people speak or seen their mouths move, Keller had not even conceived of speech. She explained, "I let her see, by putting her hand on my face, how we talked with our mouths." Sullivan used the sense of touch as a bridge between their worlds.

A *Fox Movietone News* archive clip from 1928 provides a touching portrait of the two women, seated so close to one another that they appear as one. Sullivan shows how, by lifting Keller's fingers to her face, she taught her student to communicate: "The thumb resting on the throat, right at the larynx, the first finger on the lips, the second on the nose, we found that she could feel the vibrations of the spoken word." The different finger positions enabled Keller to distinguish between letter articulations—the larynx's glottal *g*, the mouth's softer *b* and *p*, the nose's nasal *m* and *n*—a configuration so effective that it would inspire the Tadoma method of tactile lipreading. Eventually Sullivan's lessons enabled Keller to form her first sentence and, in the most poignant part of the archive footage, Keller hesitantly but clearly speaks it: "I am not dumb now." Keller experienced and connected with reality through her fingertips. Only now is science revealing that she likely did so with exceptional powers of touch.

"Neuroscientists have long known that some people have a better sense of touch than others, but the reasons for this have been mysterious," said the neuroscientist Daniel Goldreich. He and his team at McMaster University in Canada have conducted multiple studies over the past decade, surveying hundreds of people and uncovering surprising touch proficiencies. "We've shown that blind people consistently outperform people with sight in tactile acuity tests." When asked about Helen Keller, he replied, "The extreme rarity of deaf blindness means there have been few studies, so the literature on them is scant and the findings inconclusive. But I don't doubt that they too would outperform sighted people and perhaps even blind people. Keller would have had remarkable tactile acuity." His studies have also refueled an age-old battle between men and women. Conventional stereotypes would have us believe that the fairer sex is endowed with more sensitive touch. "We have repeatedly found that blind women outperform blind men in tactile acuity. They consistently score about 10 percent better than men," he said. "Also we have some evidence of this in sighted women and men." Goldreich and his colleagues wanted to confirm whether this female advantage persists throughout the population and, if so, why.

Using Scrabble-sized plastic squares that had been machine-carved with fine parallel grooves, volunteers were instructed to press their index finger against ever-narrower gaps and to identify whether the grooves ran parallel or perpendicular to their finger. Goldreich described this test "as a tactile equivalent of an optometrist's eye chart." There was a familiar pattern to the results: "On average men could sense ridges as small as 1.6 millimeters apart, but women could discern gaps 0.2 millimeters smaller still." Once again, women seemed to have the edge. "Some might assert they possess superior touch cells," said Goldreich. "Psychologists would probably argue it has something to do with the way the male and female brain work, but I suspected a far simpler solution." He decided to plot touch acuity against the size of fingertips. "We found that every person on that graph—male and female—was on the same line. This means men with small fingers display the same tactile acuity as women." So the difference is not dictated by sex. The researchers then estimated the number of Merkel cells on everyone's fingers by counting sweat pores. "Dive into that sweat pore," Goldreich told me. "At around a millimeter down, on the bottom of the epidermis, you find Merkel cells in clusters. So, these pores are a good gauge for Merkel cell density." Women have much the same number of sweat pores, and thus Merkel cells, as men, but they tend to be crowded on the smaller fingers of their smaller hands.

The star-nosed mole is living proof that cramming more sensors onto a tiny tip creates extreme touch sensitivity. Catania likened it to digital photography: "The star provides exceptional detail, much like a high-density photoreceptive chip in a camera produces a high-resolution image with many pixels." He added, "Whereas the star-nosed mole senses in single touches of short duration that create snapshots in very high resolution, we scan objects with our fingers at much lower resolution." Whether in star-nosed moles or in humans, essentially Merkel cells receive fragments of a "tactile image" and send them to the brain; the denser the Merkel sensors, the clearer the picture. A woman's touch is sensitive not because her cells are specialized; rather, their arrangements—tightly packed on typically slender hands—are more mole-like and offer a more detailed perception of the world at her fingertips. However, touch excellence is not simply skin-deep.

*

Ken Catania teamed up with a Vanderbilt colleague, the neuroscientist Jon Kaas, to investigate how the mole's brain responds to touch. He told me, "It's when I began trying to understand its brain organization and behavior that things got really surprising." They safely sedated a mole, then gently stroked the points of its star, its face, its body and watched nerves in its brain respond in real time. This activated the somatosensory cortex, a region of the neocortex associated with touch in all mammalian brains. With great care, Catania and Kaas prepared fine sections of this cortex for the microscope. They used a stain that is absorbed by activated brain tissue to illuminate the areas that had recently responded to signals from the mole's skin. They looked down the lens. "The neocortex lit up in a beautiful and complex series of pinwheels," recalled Catania, "a pattern of stain-dense stripes separated by sharp lines of low staining." These striations looked somewhat like a giant sea anemone with twenty-two tentacles fanning out over the cortical surface. The scientists realized that the staining had unveiled what appeared to be a mirror image of the mole's twenty-two-tipped star imprinted on the brain's anatomy. They could also make out shapes— alongside the star but occupying ever smaller spaces—that looked like the mole's two shovel-like forelimbs, its trunk, and whiskers. In fact, an outline of the mole's entire body seemed to be sketched across its somatosensory cortex, only distorted. The mole's star occupied around half the space, and much of this was devoted to one ray. Catania explained, "The smallest ray, number eleven, was strangely large. It took up about 25 percent of the star." The mole was depicted in the brain as a sprawling star with a tiny body in tow, recast as a star-moled nose. The discovery chimed with revolutionary research performed on people undergoing open-brain surgery without general anesthesia.

In the 1930s at the Montreal Neurological Institute, Wilder Penfield pioneered a procedure known as the awake craniotomy, used to excise brain abnormalities that were causing epilepsy. With the patient awake and talking, Penfield could observe the direct consequences of his scalpel, thereby keeping it from straying into what neurosurgeons call eloquent brain and causing irreparable damage. Such surgery is possible because our brain lacks touch receptors that register pain. The neurosurgeon needs only to administer a local anesthetic to the patient's shaved scalp before peeling back the skin, sawing an opening in their skull, and revealing its gently pulsating contents. Penfield probed over a thousand brains using an electrode that looked like an electric toothbrush but delivered weak shocks. The procedure elicited all sorts of peculiar perceptual hallucinations. Some patients smelled burned toast; others heard strains of orchestral music. Exploring the somatosensory cortex—the area where touch is processed, which runs ear to ear across the top of our head—gave rise to strange tactile sensations. Patients experienced tingles or vibrations in various parts of their body. Penfield found that exciting the same area across different brains led patients to report feelings in the same places. Marking their responses out across the cortical surface, he noticed that an extraordinary pattern became apparent. Penfield was the first to uncover a comprehensive sensory map in any brain: the first somatosensory touch map.

Traced invisibly on the surface of our brain is an eerily deformed but unmistakably human outline. It has an emaciated torso and gangling limbs overwhelmed by Brobdingnagian hands, while grotesquely vast lips and tongue dwarf its head. When Helen Keller reached for Anne Sullivan's mouth, Merkel cells in the skin of her fingers would have fired. When, in response, Sullivan touched her elbow, cells in her arms would follow. Each of these signals would have arrived in a different location of her brain, and because such information is relayed from point to point, the sensory surface of Keller's entire skin would have been represented in her brain. Scientists know that supersensitive areas of skin, dense with Merkel mechanoreceptors, are magnified in our touch map because our brains dedicate more space to process their input. The fingers take up about one hundred times more cortical real estate than the torso because they contain roughly a hundred-fold more touch receptors. Penfield called the brain's touch map the homunculus, "little man" in Latin. Over the years, as neuroscientists turned their attention to other somatosensory cortices, it was joined by a hermunculus partner and a growing menagerie: a mouseunculus, a racoonunculus, a platypunculus, among others, and recently a star-nosed moleunculus—maps that, without exception, record diverse anatomies warped through the sense of touch.

"I think of the starunculus as a fanciful drawing of the body from the brain's view," said Catania. As with the homunculus, more brain space is given to body parts that are loaded with touch sensors and where touch is paramount. Half of the mole's touch map is devoted to the star on its snout; in turn, half of that map is devoted to the tiny eleventh pair of rays. This reflects their relative importance first in finding, then in evaluating the merit of the mole's next mouthful. Catania once again saw similarities between the mole's sense of touch and image making. His thoughts mirrored those of Mark Konishi, who, on finding a map of sound space in the owl's brain, had drawn parallels with the eye and its retina. Catania elaborated, "The star's division into peripheral touch and central touch seems analogous to the retina in the visual system of mammals." The star appears to act like a retina, allowing perception of the wider landscape, whereas its tiny yet all-important eleventh rays emulate a retina's cone-dominated, small center to yield the sharpest picture. It is a tactile fovea—like the auditory one of an owl or a bat—to our visual fovea. As the star on the mole's snout presses down on the soil, it transmits to the brain a three-dimensional, star-shaped view of the terrain: a digital tactile image, with a particular focus on food. "The similarities with vision are striking," Catania concluded. "This nose may look like a hand, but it acts like an eye." The star enables its nearly blind bearer to behold the world through touch.

＊

In 2004, a man would walk into a laboratory at Harvard Medical School claiming to see as much with his fingers as most of us can with our eyes.

Eşref Armağan was born in 1953 into one of Istanbul's poorest neighborhoods, completely blind. Today, still sightless, he is a renowned artist. "I came into the world this way and I'll leave the world this way, but while I'm alive why shouldn't I learn about it?" he asked me as he sketched the outline of an apple that he could not see. "The way I learn about an object is to hold it in my hands and then to paint it." Armağan's work spans still lifes of flowers, bowls of fruit, and Turkish lutes, to portraits of accordion-playing clowns, piano players, and carpet weavers. He first feels his subjects, then transposes them to paper as an outline impressed with a sharp object such as a nail and follows the indentations with paint-slicked fingers. The final oil paintings make sense to the eye in shape and perspective, yet he sees neither the subjects nor his representation of them. The pieces have the same aesthetic and allure of naive art. Armağan told me, "Feeling my way with my fingers has completely erased my blindness. It's as if I see like anyone else."

"Sometimes nature just throws you a gift. From a neurologist's point of view a gift is not necessarily so for the patient," said Alvaro Pascual-Leone, a professor of neurology at Harvard Medical School. He and his colleagues have used an MRI scanner to investigate many unusual brains; they also used it to prove there was more truth to Armağan's claim than he could know. When Pascual-Leone began his scientific career, the prevailing wisdom held that the adult human brain is slow to change. His work is helping to rewrite this dogma. "Everything we think, feel, dream, every experience we have keeps modifying the brain," he said to me. "Rather than written in stone, our brain is dynamic and capable of rapid change." In the early 1990s, he discovered that the somatosensory homunculi of blind braille readers were different from those of people who could see; the space given to their reading fingers was notably larger. Pascual-Leone reasoned that over time, more cortical space and processing power had been allocated to the most frequently used and useful digit. "These are foveas," he explained. "We have a fovea for sight, but blind people have a fovea for touch." His words echo Catania's; the reading finger of his human subjects had become—much like the star-nosed mole's eleventh appendage—a touch fovea. Diligent everyday practice reading braille had reshaped their touch map. Such findings encouraged Pascual-Leone and others to posit that the brain has plasticity. When they scanned Armağan's, it would reveal another level of neuroplasticity entirely.

"I was scared that the machine might take something from me, but I overcame my fears, because I wanted them to take a good look at my mind, to tell me what was going on," admitted Armağan. "They put me on my back, belted my head down, and put me inside the scanner." The scientists started passing him all sorts of small objects to turn over in his hands, including a small figurine of a man sitting on a bench holding an apple. Armağan started to sketch it. "I drew it fast. It was simple. The doctors were shocked." Pascual-Leone recalled, "Eşref's skill was so good that he drew the objects we gave him with a remarkable degree of precision, in a few seconds, even lying down with the piece of paper on his belly." If Armağan's drawings were remarkable, his brain was even more so. "They next asked me to draw the man as if I was looking down on him. From the side he was long, but from above he was just a small circle on the bench." Pascual-Leone explained, "A different point of view is an intrinsically visual thing; there is no tactile point of view. What I found striking was that he has vantage point and perspective without ever having seen. It suggests that this spatial awareness is somehow ingrained, innate."

In order to ascertain what part of Armağan's brain is activated when he translates the world he touches into drawing, the scientists asked him to handle the same objects but then make unconnected doodles. This enabled them to separate the input of handling from the output of drawing and depicting. They observed that the act of touching predictably would excite the blind artist's somatosensory cortex and his touch fovea. Furthermore, the process of drawing, as opposed to doodling, would activate three other areas of the brain: the prefrontal cortex, the lateral occipital cortex, and the visual striate cortex. "Eşref is congenitally blind, but if you were to see this picture of the brain, not knowing what is going on, you would say this person is seeing," said Pascual-Leone. The primary visual cortex, found near the occipital protuberance—the bump at the back of our skull—is traditionally where sight is processed. In the absence of visual input, Armağan's sense of touch not only fired his somatosensory map but had also co-opted a visual part of his brain.

Armağan was thrilled with the discovery, believing it confirms what he had always hoped: "I am not blind, I can see. I was the first person to prove that you don't need eyes to see." He is not alone in equating the senses of sight and touch. John Hull, whom we met in the previous chapter, expressively captured the experience of slowly going blind and wrote, "I am developing the art of gazing with my hands. I like to hold and rehold and go on holding a beautiful object, absorbing every aspect of it. . . . The blind person sees with his fingers." Pascual-Leone wonders whether there is some truth to the idea—quoting Picasso's famous line, "Painting is a blind man's profession"—but the question is, What truth? "I think that the characteristics of Eşref's internal representation are fundamentally the same as in any of us," he said. "At the very least, Eşref can convey a representation of reality that is inside him in such a way that he translates it into an image we all unequivocally recognize." But exactly how? Just because Eşref assimilates the circuitry typically used for sight, is that seeing? Such questions are not mere semantics; they expose the gaps in our knowledge of what it is to see. The scans show how Armağan's lack of sight has improved his sense of touch by co-opting the extra power of his visual cortex, a phenomenon that is not solely the preserve of blind people.

*

"I sometimes wonder if the hand is not more sensitive to the beauties of sculpture than the eye," wrote Helen Keller. "I should think the wonderful rhythmical flow of lines and curves could be more subtly felt than seen." Keller used her fingers both to see and, as they flickered over Anne Sullivan's face, to hear. Daniel Goldreich points out that deaf people have also been shown to possess superior touch. Similarly, their sense of touch takes over the unused auditory cortex, where hearing is typically processed. Without sight and without hearing, areas of Keller's brain were freed up to be used elsewhere. Goldreich explained, "It is entirely reasonable to suggest that Keller's touch, as well as taking advantage of homunculus somatosensory neuroplasticity, might also have been making use of both her visual and auditory cortex." Pascual-Leone concurred: "A scan of Keller's brain would be fascinating. Her visual cortex must surely have been taken over, and perhaps her auditory cortex too." This adds up to much more brainpower being devoted to such a person's sense of touch. Neuroscience has shown how people who are blind or deaf are supertouchers. Arguably, the same logic would make deaf-blind people superlative supertouchers.

Pascual-Leone believes we all share this promise. To prove it, he undertook a highly ambitious study by asking people for permission to make them temporarily blind in exchange for an unforgettable experience. For five days, the brave volunteers were confined to a hospital bed and blindfolded. They could not see any light whatsoever; they were absolutely without sight. "The blindfold contained photographic paper, so if they tampered with it, we'd be able to tell." Not one person broke the pact, even when the experience became grueling. By day two, they were suffering from visual hallucinations. "People would go into the bathroom to wash their face and claim to see their face in the mirror. You might ask how they could see with a blindfold. They said that the image they saw of themselves in the mirror was wearing a blindfold!" Some saw Lilliputians or cartoon figures; others saw complex cityscapes, beautiful skies, and sunsets. "This is what a brain does," Pascual-Leone told me. "It makes up stories; it conjures up imagery. Our brain plays with us continually." The subjects whiled away the hours learning braille. Once a day, they were wheeled to a PET scanner and asked to perform simple touch tasks as it mapped their brains. Pascual-Leone discovered two things. First, "it turns out that being blindfolded both day and night for five days is what it takes to learn braille." The second revelation was far greater.

By day two, the scans showed that the volunteers' brains were already starting to change. By day five, the same brains were behaving much like Armağan's had: the act of a touch stimulated their visual cortex. "They were seeing with the tip of their finger," said Pascual-Leone, "but what surprised me was the speed in which it happened; I was expecting changes, but not this rapid. Given we know that neurons grow only millimeters per day, it would take months, even years to build new connections, so something else was going on." He surmises that the connections must already have been there and that the experiment simply unmasked them—evidence not merely in favor of exceptional neuroplasticity but of a whole new way of looking at the brain. "It's provocative, but we're arguing that the brain may not be organized into sensory modalities at all," he said. "The striate cortex is visual only if you have vision. If you don't, it quickly takes on other sensory modalities." What neuroscientists have called the visual cortex for the past century seems not to be devoted exclusively to the eyes. Pascual-Leone wonders whether it might more accurately be defined as the area of the brain best able to discriminate spatial relationships and that it will use any relevant sensory input. Usually this is information coming from the rods and cones of our eyes, but equally it could come from the hair cells of our ears or the mechanoreceptors of our skin. "We are still a long way from knowing how the brain works," said Pascual-Leone, quoting the late, eminent neuroscientist Paul Bach-y-Rita. "You can do so much more with your brain than Mother Nature does with it."

✳

The star-nosed mole epitomizes nature's already astonishing versatility, teaching us much about our sense of touch through an organ we normally associate with our sense of smell. Its star-shaped nose may be the most sensitive appendage among mammals, but it is primed with the same cells found in our skin and at our fingertips: sensors that bring us the feel of the world beyond. Famously, Bach-y-Rita also said that we see not with our eyes but with our brains. Similarly, we do not solely hear with our ears, smell with our noses, taste with our tongues, or feel with the sensors in our fingers. Even by this line of logic—with our brain as our body's all-important sense organ—the mole confounds expectation. The way its star is etched across its brain means that, in practice, its nose, which looks like a hand, acts more like an eye. Ken Catania told me,

> Some might suggest that the mole's visual cortex was taken over by touch, but there is no line in the brain that says, "Here is where the visual cortex should be." Think of it another way: somewhere there is a star-nosed mole writing a book on humans and wondering if our strangely large eyes took over all the area that is supposed to be for touch.

In *De Anima*, Aristotle declared, "Man is inferior to many of the animals, but in delicacy of touch he is far superior to the rest." Clearly, he was unacquainted with our stellar mole.

5

The Common Vampire Bat and Our Senses of Pleasure and Pain

Dusk falls over the jungles of Costa Rica, shadows stretch, and a vast colony of bats roosting in the murky recesses of a cave surfaces from a day of inactivity. The wing of one extends, and its neighbors follow. Movements ripple out in concentric circles over the ceiling. Slowly they emerge from the cavern mouth, gathering into skeins that streak the darkening sky. The morning of the vampire bat, *Desmodus rotundus*, has begun. With a throat too narrow to swallow solid food, it survives solely on a liquid diet of blood. It is the only mammalian sanguivore known to science. A pair of razor-edged upper incisors can puncture the tough hides of horses and cows with ease. Then, over the course of twenty to thirty minutes, the bat sips leisurely at their blood through two straw-shaped ducts on the underside of its tongue. Its saliva holds as many drugs as a pharmacist's cupboard: proteins with antibacterial and antimicrobial properties; a vasodilator to expand capillaries and increase their flow; an anticoagulant ghoulishly named Draculin to prevent clotting and ensure a continuous trickle. No wonder such an adept bloodsucker stoked the imagination of a certain Victorian gothic novelist. Once immortalized as the alter ego of the diabolical Count Dracula, the bat became the embodiment of evil, condemned to slip through the darkness in search of unconscious prey. However, one scientist would find the vampire bat much maligned.

"It all started with a comment at a conference from a distinguished professor who had been raising orphaned vampire bat pups in captivity," said Gerald, or Jerry, Wilkinson. It was 1978, and Wilkinson, a recently graduated zoologist, was in Costa Rica under the auspices of the University of California at San Diego. The professor to whom he was referring was Uwe Schmidt, founder of a captive bat colony in Germany. The comment concerned some odd behavior: adult bats were known to regurgitate ingested blood to feed their own pups, but Schmidt had seen them doing so with orphans. Mothers sharing food with their own children is commonplace in nature, and it makes sense to wean pups from milk to blood before they fly off on their first hunt. However, to extend this nurture beyond one's own family—to give blood to nonblood—was unheard of. Wilkinson added, "Food sharing between nonfamily members of any species ran counter to the dogma at the time." The gene-centered view of evolution, which espouses ruthlessly competitive individualism, was gaining ground. Richard Dawkins had only recently popularized the theory in his book *The Selfish Gene.* Moreover, a study had shown that it takes only two nights without food for a bat to starve. "Vampire bats live on the edge," Wilkinson explained. "Blood is a lousy source of energy. It contains very little fat, so vampire bats do not have the energy stores of other bats or most mammals. Those that fail to feed for sixty hours can lose a quarter of their body weight and become unable to maintain a critical body temperature." So to share such hard-won sustenance with nonrelatives seemed less likely still. By good fortune, Wilkinson was already uniquely positioned to satisfy his curiosity.

He was staying in the northwest of Costa Rica at a cattle ranch called Hacienda La Pacifica. He had noticed that its livestock were an irresistible draw to a vigorous community of vampires. Following the bats, he found that in the absence of caves, they were roosting in large hollowed-out trees. Wilkinson's first challenge was to catch his subjects and tag them with identifying bands. "We're talking months of severe sleep deprivation," he told me. "People think bats come out with the moon, but they avoid bright light and wait until the moon has gone down. So depending on the moon cycle, we had to be at the roosts at various early hours in the morning." On some days, he and his team would net, then weigh and tag bats leaving the trees; on others, they would catch them on their return. He was struck by how many were flying home on empty stomachs. "It was easy to spot. You don't need to weigh these creatures to know whether they have eaten. They consume between 50 to 100 percent of their body weight in blood at one sitting and, like ticks, their bellies can swell up to twice their size." The netting studies showed that on any given night, as many as 7 percent of the adults and 30 percent of the young fly home hungry. Wilkinson computed that such failure rates ought to result in 80 percent mortality. Yet the colonies were thriving. The solution to this conundrum would involve doing something not to everyone's taste. Wilkinson and his team would be the first to study vampire bats in the wild, up close.

To access the dark, cramped roosts, the scientists had to lie down on their backs and wriggle through the small openings in the trees. "I often had researchers come out screaming," Wilkinson admitted. It turned out that the close proximity of a cluster of vampires was nothing compared to that of the giant cockroaches. "Our lights would pick out these 3-inch [7-centimeter] roaches scuttling up inside the trunks and their nymphs infested the mulch we had to lie on." The scientists worked in pairs—taking turns being "the eyes" observing or "the pen" noting—giving each other invaluable moral support over the long, confined hours. Their determination paid off. "To begin with, the bats hid from our lamplights, but as time went on, they became accustomed to us and started to behave the same with or without the light on." That year, the team saw all manner of behaviors never seen before in wild vampire bats: females fighting, giving birth, and nursing their own young. Just as Wilkinson was giving up hope of spying the strange captive bat conduct that had kept him in Costa Rica, his luck changed.

"At the start, we didn't always know who was who, but as we figured out each colony's genealogy and got to know who we were looking at, we saw it." To recall the exact date and times, Wilkinson had to dig out the expedition's meticulously detailed logbooks. "It was the morning of July 28, 1980," he told me. They were watching two adult females locked together, wings wrapped in a leathery embrace. "At 11:09, we saw one start licking the other's mouth—consistently for over a minute. We realized it was licking up blood that the other had just regurgitated." Another logbook entry from the following morning reminded him: "That's when we saw another adult female regurgitate blood for a pup that was not her own." These were the first observations in nature of nonfamily sharing of blood, not only across generations but now also between adults. Once the team knew what to look for, they saw it repeatedly. They also started to notice that the behavior followed elaborate sequences of grooming. "Typically, a bat would lick another, perhaps under its wing, but when it started licking the other's lips, this would often prompt the bat being groomed to regurgitate." The team had already noted the bats spending inordinate amounts of time licking, scratching, and nibbling one another, but now it became apparent that these rituals were being used to encourage sharing. "In the 600 hours we collectively spent lying on our backs over those five years, we witnessed 110 instances of bats feeding other bats, mostly between a mother and her pup, but 30 percent involved adult bats feeding one another or young that weren't their own." The conclusion was incontrovertible: that casual comment at the conference had been not simply validated but also found commonplace.

Evolutionary biologists leaped on the study as a rare case of selflessness at work among selfish genes: an example of reciprocal altruism that exists because there is a trade of favors. A bat who has flown home on a full stomach donates life-saving blood to another in exchange for being groomed and the likelihood that one day the tables will be turned. As the neuroscientist David Linden quipped in *Touch: The Science of Hand, Heart, and Mind,* "I'll groom you, and you'll vomit blood down my throat. Next time, in the spirit of reciprocity, maybe I'll do the same for you." Richard Dawkins included the study in the next edition of *The Selfish Gene.* Vampires, he wrote, "could herald the benignant idea that, even with selfish genes at the helm, nice guys can finish first." David Attenborough and a BBC crew were the first to capture the behavior on film for their landmark nature documentary series *The Trials of Life.* At a stroke, Wilkinson had rewritten both evolutionary textbooks and the vampire bat's reputation: Dracula's children are not the malevolent monsters of legend, but caring and sharing creatures. Their honorable code of conduct is mediated through hours and hours of social grooming. Altruism arises between the give and take of a touch. Bats are not the only species to wield and succumb to contact. Indeed the most momentary brush of skin on skin can make us all behave in surprising ways.

*

On a sweltering summer's day in 2004, in the medieval walled town of Vannes on the French Breton coast, thirsty customers were sitting at a local bar waiting to be served. They were unaware that they would shortly be taking part in an experiment investigating the power of human touch. The waitress on duty that day was under strict instructions to touch half of her customers on their arm, surreptitiously, when taking their orders. Tipping at bars in France is not obligatory because a service charge is included in the bill. On that day, the researchers discovered that a small, seemingly insignificant touch was enough to prompt more customers to leave an extra *pourboire* for the waitress. Elsewhere, scientists have similarly found that a casual touch improves the chances of a bus driver giving someone a free ride, of students participating in class, of a smoker acquiescing to a request for a cigarette. Moreover, add touch to a transaction, and people tend to return a dime they just found in a phone booth to its rightful owner or report feeling more satisfied by a pitch from a secondhand-car salesman. In many of these studies, the swayed subject had not even noticed being touched. Fleeting contact between strangers is mysteriously potent: capable of inducing generosity with money, time and effort, honesty, and even happiness. The phenomenon has been called the "Midas touch," after the mythological Greek king whose touch turned all to gold. A whisper of an explanation can be heard among our everyday conversations.

"Why are emotions called feelings and not sightings or smellings?" asked Linden. "This may seem like a silly question but it's not." Touch metaphors suffuse language, as Diane Ackerman noted so vividly in *A Natural History of the Senses*: "We care most deeply when something 'touches' us. Problems can be thorny, ticklish, sticky, or need to be handled with kid gloves. Touchy people, especially if they are coarse, really get on our nerves." We call someone who is emotionally insensitive "tactless," lacking touch. We describe being spurned in love as "hurtful," resulting perhaps in "heartbreak." The physiologist Frederick Sachs, pointing out that touch is the last of our senses to fizzle out, wrote, "Long after our eyes betray us, our hands remain faithful to the world . . . [and] in describing such final departures, we often talk of losing touch." The concepts of touch and emotion are entwined in languages around the world for good reason: there is an emotional frisson to touching and being touched. As Ackerman put it, even an encounter "so subtle as to be overlooked doesn't go unnoticed by the subterranean mind." But the palpable embrace of a friend, a mother's tender kiss, or a lover's lingering caress all readily release floods of feelings. No other sense arouses us so strongly. We are touch-hungry creatures, touch-starved with frightening ease. Our understanding of the importance of touch is changing the way in which we define this sense and slowly starting a scientific revolution.

In 2014, a paper published in the scientific journal *Neuron* bemoaned the fact that nearly all research of the past hundred years, or thereabouts, has focused on our hands and what the authors called "discriminative" touch. This sense enables us to grasp the lay of the land: to explore its size, shape, and texture. It is how the star-nosed mole feels its way so expertly through underground burrows and, as the previous chapter established, how touch receptors—Merkel, Meissner, Pacinian, and Ruffini cells—play subtly different and complementary roles in building perception of the world at our fingertips. However, Francis McGlone, a professor of neuroscience at Liverpool John Moores University and the lead author of the *Neuron* paper, argued that research has more or less ignored another touch system in our skin. Consequently, he told me that of all Aristotle's five primal senses, touch is the least understood: "Touch is our last great sensory frontier. We are learning that it is an inadequate word for the complex senses it encompasses. Historical scientific inquiry may have concentrated on the digits of the hand, but touch is sensed across the whole body." Now scientists need to focus their efforts on the previously ignored expanses of skin that cover our limbs, back, and chest. "There is mounting evidence that touch has another dimension, providing not only its well-recognized discriminative input but also a social and emotional side. We are only now uncovering these circuits and systems." In much the same way as the cones and rods of our eye divide vision into daytime color and night sight, it seems touch diverges into at least two senses: one with an emphasis on *touching*, the other on *being* touched.

As our hands explore an object, mechanoreceptors send electrical neural impulses to the brain at superfast speeds through fat-insulated nerve motorways. Our skin also contains slender fibers, with little to no insulation, and these country lane nerves take a leisurely pace. "We have a rapid first touch system that fires over 100 miles [160 kilometers] per hour and reaches the brain in milliseconds," said McGlone, "but we also have a slow second touch system that follows at a couple of miles per hour and arrives a second or two later. About three-quarters of the touch nerves in our skin belong to this second system." In this form of touch, the nerve cells themselves are the sensory receptors: long peripheral neurons that link skin to spinal cord. Their microscopic bare endings are spread throughout our epidermis. They gather deeper in our dermal layers, like a tree's root system into a nerve trunk, from limb, belly, back, shoulder, and scalp to the spine, where they excite other nerves venturing brain bound. Ultimately, whereas our skin receptors with their fast fibers relay the topographic detail beneath our fingers to our touch fovea with near-instant gratification, there is mounting evidence that these more prevalent slow sensory fibers build the emotional tone of a touch.

McGlone is part of a small but committed global team of neuroscientists, neurophysiologists, clinicians, and, among others, neurologists whose collective attentions have become centered on one particular slow sensory neuron found in our skin only relatively recently. "It has been my passion for the last twenty years, since reading a paper by Åke Vallbo, the man who discovered it in humans," McGlone told me. "Finding it has meant we have had to work out what it is for." Studies have shown that it is absent from the hairless skin on the soles of our feet, the palms of our hands, our fingertips and lips. "Absent from the skin that uses discriminative touch to explore the world," he explained. "This anatomy tells us something; it gives us a clue to its function." Scientists have recorded how the neuron responds to all manner of prods, pokes, and strokes. "We have found that it only reacts to very low forces, anything below five-thousandths of a newton. Only to low-velocity movements around 3 to 5 centimeters [1 to 2 inches] per second. And does so optimally when the touching stimulus is at skin temperature"—in other words, to the same pressure, speed, and skin-on-skin warmth of a caress. "This makes sense; the hairy parts of our skin are, after all, where we generally like to be stroked or hugged." Moreover, brain scans have shown that whereas discriminative touch excites our somatosensory cortex, these neurons target the parts of our brain that process emotion, such as the insular cortex and limbic structures. The combined evidence indicates that this sensor is uniquely primed to the caring touch of another human being.

The sensor has various names: the C-low threshold mechanoreceptor, C-tactile, CT fiber, or, according to David Linden, the "caress sensor." McGlone said, "I have toyed with terms *hedonoceptor* and *hedonoception*," from the Greek Hedone, the goddess of pleasure, for our pleasure sensor and our sense of pleasure. "Truth be told, we wrestle with the language as, strictly speaking, its activation creates perceptions that aren't exactly ecstatically pleasurable so much as mildly pleasant." Either way, it registers touch in a way that we find rewarding. McGlone and a colleague even redrew Wilder Penfield's touch map of the brain, accentuating body parts where it most densely innervates the skin to make what they described as a "hedonic homunculus." Its outline is no less grotesque than our somatosensory homunculus but differently misshapen: our hands, lips, and tongue have shrunk back to proper proportion, but our shoulders, upper arms, scalp, and back are now monstrously swollen. "We have also found it in every social mammal that we have so far looked at," McGlone said. "In evolution, animals that work together are more successful; to encourage togetherness, there was a need to reward close physical contact. This sensor does that, and we believe it plays a fundamental role in nurture, bonding, and social touch." In all likelihood, this touch nerve explains why a vampire bat gifts blood after it has been licked, why humans yield to the lightest of touches, and why we turn to touch metaphors to describe emotion. By tuning us to tenderness, it transforms touch into interpersonal glue and the skin into a social organ. "Essentially its role is not to sense the physical world but to feel it," concluded McGlone. The pleasure sensor is not the only neuron within our skin that "feels" as opposed to gathering facts. Like the many receptors of discriminative fast touch, this company of neurons works together to build the complex and subtle mood of an encounter. It runs the full gamut of emotions, from pleasure to pain.

✳

"Have you ever been out in the bitter, bitter cold, where your feet were ice, almost like frostbite? Then you warm them up and it burns?" Pam Costa, a professor in psychology at Tacoma Community College in Washington State, asked. "I feel that overpowering burning sensation even though it might be cold outside." To explain, she recalled an occasion when she was doing some housework. "The iron was burning the inside of my wrist, only I didn't know it until I heard and smelt my skin searing. It wasn't that I couldn't feel, but because my hands often feel like they are being burned, I didn't feel the difference." Such experiences are an everyday reality for Costa. She has to wear loose clothes because the fabric against her skin can feel like a blowtorch. Even the most mundane acts—putting on shoes or entering a warm room—can trigger an attack. When she was a young child, doctors struggled to diagnose what was wrong with her. Given that no objective measurement of pain exists—patients are simply asked to rate its intensity on a scale from zero (no pain) to ten (unbearable)—she was repeatedly told that her torment was imagined. "There was one highly regarded doctor, a university head of rheumatology, who took three days giving me all sorts of tests. I was really hopeful he would help, but he told me that my pain was entirely psychosomatic and that I needed antidepressants. That was probably my lowest moment." It was not until 1976, when Costa turned eleven, that a letter arrived, addressed to her parents, from the Mayo Clinic. She learned that she was not alone.

Costa discovered that she suffers from an unusual condition first observed in the nineteenth century on the battlefields of the American

Civil War. The physician Silas Weir Mitchell saw that some wounded soldiers complained of pain, albeit in the absence of any visible cause. He diagnosed nerve damage and named the neuropathy erythromelalgia, from the Greek words for red, limb, and pain—*eruthrós, mélos*, and *álgos*. Sufferers say they feel as though they are on fire or being scalded, and some liken it to being touched by volcanic lava. And so the syndrome has become known simply as "Burning Man" or "Man on Fire." The letter from the clinic informed Costa that, to her astonishment, there were twenty-eight other members of her wider family with Burning Man, across five generations, all of whom came from Birmingham, Alabama. "They are cousins on my maternal side," Costa told me, "but because they lived in Alabama and we were in California, I didn't know of them." Their existence was the first evidence that this rare condition can be inherited. They piqued the interest of a neurologist whose mission in life is to find a cure for pain.

"Chronic pain affects more than 250 million people worldwide. It occurs more frequently than cancer, heart disease, and diabetes combined," said Steve Waxman, founder and director of the Center for Neuroscience and Regeneration Research at the Yale School of Medicine. "The pain medications we have are often ineffective and can produce incapacitating side effects, like confusion, loss of balance, sleepiness, even addiction. To cure chronic pain, we need to understand where it comes from." More than a century ago, the eminent English neurophysiologist and Nobel laureate Sir Charles Sherrington was the first to set out pain as a specific sense. He coined the terms *nociception* and *nociceptors*—from the Latin verb *nocere*, to hurt—for our sense of pain and its sensory neurons. As with our pleasure sensors, the unadorned endings of these pain neurons are rooted in our epidermis. They are also present in our gut, muscles, joints, and wherever else we ache, sting, or smart. If we touch a flame, we feel a sharp, instant pain that makes us withdraw. This subsides quickly, but the chronic, burning sensation that soon follows can last for hours and carries an emotional tinge. Like pleasure, our sense of pain is more about feeling than fact. Sherrington had no way of knowing, but not only are most nociceptors from the same slow touch system as hedonoceptors but the two are also neurologically connected. Although more is known about our pain than our pleasure sensors, Waxman had his work cut out because how they work remains largely mysterious. "There are more than 100 billion nerve cells in the human nervous system, greater than the number of stars in the Milky Way, and millions of these are nociceptors," he told me. If our pain neurons were to light up, our bodies would resolve into an intricate and crowded cosmic web. Waxman had an idea to help narrow down the investigation. The blueprints to these nerves—along with the rest of us—are contained in the human genome. "I decided to search for a pain gene. By that I mean a gene that encodes protein molecules that are central players in pain. The discovery of the extensive Alabama family was immensely exciting as their chromosomes were the perfect place to start looking." With blood that had been extracted from Costa and her kin, Waxman began to hunt for the single gene among the twenty-odd-thousand that was causing their anguish. Meanwhile, in northern Pakistan, rumors surfaced of a case that would prove invaluable to Waxman's quest, described by one geneticist as the "strangest to appear in scientific literature."

*

Crowds gathered on the streets of Lahore to watch a young boy act out feats that defied belief. With great bravado, he pierced his flesh with knives and walked barefoot across red-hot coals. People praised his fearlessness, but he claimed to be immune to pain, unable to feel even the slightest twinge. A life without suffering sounds appealing and probably like nirvana to those afflicted with Burning Man, but in reality it can be more of a curse than a blessing. Pain is our guardian angel. It warns us of the world's many and often invisible dangers. Its various guises—whether slicing stab, grip-like vise, or dull ache—enable us to discern scratches from stings, burns or bruises from broken bones. Pain prompts us to stop walking on sprains, pull away from scalding water, or perhaps call the doctor. Its absence might fool us into feeling invincible. On his fourteenth birthday, the street performer decided to impress friends by jumping off the roof of a house. The decision killed him. The tragedy underscores how vital pain is for survival. It also caught the attention of a team of British scientists.

Researchers at Addenbrooke's Hospital in Cambridge wondered whether the boy had been affected by yet another rare inherited condition known as congenital insensitivity to pain. They decided to track down his relatives. Sure enough, they found six other children, ranging in age from four to fourteen, from three families of the northern Pakistan Qureshi clan. All were covered in cuts and bruises, with self-inflicted injuries to their lips. Some were missing the tips of their tongues, having bitten them off; others limped on unhealed fractures. Under neurological examination, they could feel the touch of a needle, the tickle of cotton wool, the pressure of fingers squeezing their arms— proving that their sense of discriminative touch was unimpaired. Yet they had no experience of pain. More vigorous prods and pokes, even blood tests, elicited no response whatsoever. As the young Lahore street performer had boasted, none of them knew what pain felt like. The symptoms and the family pattern confirmed the diagnosis of congenital insensitivity to pain. This time, blood samples offered a unique opportunity to hunt down a gene that might erase pain.

At the same time, Steve Waxman and the Yale team were making headway. They had homed in on a portion of chromosome 2, the second largest of our twenty-three pairs. Waxman explained, "The result was important; it had limited the search from twenty-odd-thousand genes to about fifty. We no longer needed to find a needle in a haystack. Now we could focus on a small but still formidable tangle of hay." Then Waxman saw an article in the *Journal of Medical Genetics* of a study conducted on a Chinese family with high incidences of Burning Man. A Beijing laboratory had found that all the patients had mutations of a gene known as SCN9A. "When I saw the title, I told my team that it was a bad day and retreated to my office; it appeared at first blush that we had been scooped," he recalled:

It was only after a cup of black coffee and a thorough reading of the article that I realized what had happened. Finding the mutation did not prove that it caused the patients' pain. We still had to show how the SCN9A gene changes the function of pain neurons in a way that explains the disease.

He set his team the new challenge of ascertaining how mutations of this gene make patients like Pam Costa feel as though they are on fire. Then Waxman learned that the team at Addenbrooke's had also narrowed their search to chromosome 2 and, extraordinarily, the very same SCN9A gene. "I was overwhelmed when I saw both sides of the genetic coin," Waxman explained. "That was when I realized that SCN9A really could be a master gene for pain." Different mutations were causing antithetical symptoms; the Alabama one makes cool surfaces feel like red-hot coals, but the Pakistani variant means the coals are not felt at all. Efforts were redoubled to understand how this gene affects the pain neuron, to either inflame or extinguish experience. "What had appeared to be a bad day was, in fact, a good day," said Waxman. "There was a lot of work to do. The ball was now in our court." The challenge that lay ahead remained colossal. Our sense of pain is highly complex. It also conflates sensory systems one might not first associate with pain, as exemplified by our crepuscular bloodsucker.

*

Few could have guessed what lurked behind a locked door within the Poppelsdorfer Schloss in the German city of Bonn, but every week buckets arrived from the local slaughterhouse brimming with fresh blood. The Schloss was home to both the University of Bonn's Zoology Department and a thriving community of vampire bats started by Uwe Schmidt in 1970. Schmidt, of course, was the esteemed professor whose tales of caring vampires had spurred Jerry Wilkinson on his adventures. He had been studying rabies in Mexico. Vampires do not often bite humans, but if they do, they can pass on the lethal virus and indeed are responsible for most cases of human rabies. Schmidt found himself falling under their spell. So when it came to boarding his flight back home, he decided on unconventional company. "I built a special transport box—thirty compartments for thirty vampire bats—and to the surprise of the cabin crew, I carried it with me as hand luggage," he told me. The Schloss's new arrivals would form the first and only captive vampire bat colony beyond American shores: the closest that living, breathing, and breeding vampires ever got to Count Dracula's Transylvanian castle.

The bats settled into their new home, and over the years, thirty multiplied to more than one hundred. Although Schmidt recruited a team of zoologists to care for and feed them—"easily getting through 2 liters [3.5 pints] of blood a day"—he prided himself on being able to recognize nearly every bat. "Vampire bats are very misunderstood," he told me. "In reality, they are lovely, affectionate animals and the most intelligent of bats, with the largest neocortex." One in particular stood out: "Pipifax was my favorite. She was very handsome and because she was born in captivity, she became so tame I could tuck her inside my shirt pocket and walk her around the laboratory." Another female lived for twenty-nine years, bore many pups, and became the colony's undisputed matriarch. She, Pipifax, and the various offspring would afford the scientists many insights into nature's vampires, including a rare form of touch that does not even require physical contact.

Schmidt was intrigued by how bats target their victims on dark nights, when the moon has set and slipped beneath the horizon. "Very little was known about how vampire bats search for prey," he explained. "I wondered if they could sense subtle differences in temperature. This would be a great advantage to an animal feeding on other warm-blooded animals." The ability to feel temperature is another sense of touch. This thermoception—like hedonoception and nociception— relies mainly on slow-firing neurons. Whereas we can warm our hands against a roaring fire, we cannot sense one another's body heat unless we are touching. To do so from a distance is a rare talent. "It had been shown only once before in vertebrates," said Schmidt, "in an important study of cold-blooded pit vipers blindly attacking warm-blooded mice." He enrolled his graduate student Ludwig Kürten and asked him to investigate this possibility in the captive vampires.

"It became the subject of my master's thesis," Kürten told me, "so I was keen to design a good experiment." He built two units with heating elements on copper plates and mounted them next to one another on the far wall of a cage: "Leaving one unit at room temperature, we heated the other up to 37 degrees centigrade [98.6 degrees Fahrenheit] to mimic the bat's warm-blooded prey and hid a tube of blood beneath it." With the lights turned off and the tube hermetically sealed to ensure that the blood could neither be seen nor smelled, he released two bats into the cage, one after the other. Vampire bats are unusually accomplished walkers, even runners. "Each bat crawled across the cage and started to investigate the two units," recalled Schmidt. Once it had realized the warm unit held the reward of blood, the scientists then forced the bat into making a decision that would confirm whether it could sense body heat remotely. They placed a board between the two units. "We started with an 8-centimeter-deep [3 inches] division, so as each bat approached, it had to choose which unit to head toward." The scientists waited and watched; reliably, the bats turned toward the warmer of the two units. They deepened the division again and again; still the bats chose the option with the treat even from 16 centimeters (6 inches) away. The experiment had proved that vampire bats have the rare sensory capacity to feel heat from a distance. Schmidt and Kürten called it "thermoperception." Their findings were published to great excitement within the scientific community, but Kürten's overriding emotion was one of relief: "I was just glad that the experiment had worked." The next challenge was to locate the vampire's sensory organ.

Filming the bats' faces with a thermal-imaging camera was instantly revealing. "They looked so funny," remembered Schmidt. "Each bat's face appeared bright red, with a very blue nose-leaf in the center." The nose-leaf is a structure common to many bats and to all vampires. It lies on top of dense connective tissue so protrudes like a pig snout and, thermally insulated, is 9 degrees centigrade (48.2 degrees Fahrenheit) cooler than the rest of the bat's body. "The thermal organ had to be located in this area," Schmidt told me, since only then could the bat distinguish the prey's body heat from its own. The vampire's nose-leaf is surrounded by three pits. "To begin with, we thought we would find the heat receptors in the leaf-pits; their hairless and glandless structure made them ideal. They had been found before in the pit organs of crotalid snakes," said Kürten, "but we turned out to be wrong." The scientists used a technique called electrophysiology, which eavesdrops on a nerve's electrical chatter, but their examinations of the pits drew a blank. "There was no response to heat in any of them." Instead, they found heat responses in the nose-leaf itself, particularly along its rim. According to Schmidt, "People still argue mistakenly that the pits are the vampire's heat organs, but it is very clear that it is the creature's actual nose-leaf." Further analysis of individual sensory neurons highlighted such fine-tuning to the temperature of warm-blooded bodies that the bat can sense the heat of its victim's veins. "A vampire bat tends to use echolocation to locate its prey in the dark, and its hearing is particularly sensitive to the faint sound of breathing," Schmidt told me. "Thermoperception takes over when it comes in to land. The bat uses this sense as it crawls over its prey to survey the skin and bite where the blood flows closest to the surface." This heat-seeking biology makes the vampire bat unique among mammals, but the chemistry behind it is not so distinctive.

*

"The work of the Schmidt Lab on the bat behavior and anatomy was superb, but no one had looked at the cellular physiology—the molecular logic," explained David Julius, a biochemist at the University of California, San Francisco. He set his laboratory the task of doing just that. Researchers began by dividing up the nerves of a bat into those above its neck, which innervate the face, known technically as the trigeminal ganglia, and those below the neck, which innervate the body, the dorsal root ganglia. "Our theory was that the nose-leaf, as part of the trigeminal system, would be doing something different from the rest of the body. We wanted to find out how this was reflected molecularly," Julius told me. They dissolved the two sets of nerve cells and extracted their DNA. At first, they could not find any difference; the genes expressed in the nerves of the bat's nose-leaf were the same as those in its furry body. A closer look indicated that one of the proteins encoded by these genes appeared in two forms. "We realized the bat was splicing a certain protein. We found a long version in the dorsal root ganglia and a much shorter, spliced version in the trigeminal ganglia." The regular long form primes the neurons in the bat's body to temperatures over 40 degrees (104 degrees Fahrenheit) that might harm it; the shortened nose-leaf version senses temperatures almost 15 degrees (59 degrees Fahrenheit) lower. The Julius Lab had found the specific molecule that enables vampires to feel the presence of another from only their faint glow of heat. It is a short protein with a long name and an awkward acronym, a thermo-transient receptor potential, or TRP ("trip") V1. This vampire warmth-sensing chemistry is found in every one of us.

The bare touch nerve endings in our skin are peppered with TRP proteins. As in the bat's body, TRPV1 primes us to feel the blistering temperatures of boiling water or smoldering embers. We perceive these temperatures as painful and pull away. The Julius Lab—which investigates all manner of pain-inducing substances, from the spiciest chili pepper to the skin toxins of the yellow-bellied toad—deduced that the same TRPV1 molecules would fire if we were unfortunate enough to be bitten by a fast and flighty Trinidad chevron tarantula. They are also affected if we lounge for too long under blazing summer skies and get sunburned. These nerve proteins react with the same burning sensations to bites, stings, scalding water, and searing coals. Consequently, they blur the distinction between our senses of temperature and pain. Moreover, we know now the TRPV1 sensor is but one example in an expansive family. Some have been found to fire at even higher and more painful temperatures; others respond to balmy and gentle warmth, and still others to coolness. Ultimately, scientists like Julius expect to uncover many variants, acting together to endow a seamless perception of temperature alongside a sliding scale of pain. Meanwhile, research is suggesting that they work in tandem with the component misfiring in people with either Burning Man or chronic insensitivity to pain.

✳

Steve Waxman had pressed on with his mission to understand the pain gene with renewed fervor. "There was a team of thirty-five scientists in my lab working round the clock." Now they knew that the SCN9A gene encoded for yet another protein: a sodium channel within the pain neuron's membrane. In laboratory petri dishes, they created nerve cells like those in Burning Man patients by inserting the mutant channel into living tissue. Then, like Uwe Schmidt and Ludwig Kürten, they used electrophysiology to monitor and measure these Burning Man neurons in action, cell by individual cell. "The results were dramatic," said Waxman. "The mutant channels shifted their activation by 13 to 14 millivolts. One millivolt is one-thousandth of a volt; that may not seem large, but from the point of view of a neuron, it is huge." It was evidence that the Burning Man pain sensors were far more likely to fire and to do so at abnormally high frequencies. "We could see that the pain-signaling neurons were hyperexcitable," he explained. "They were screaming when they should be whispering." By contrast, the SCN9A mutation found in the pain-insensitive Pakistani families has the obverse effect: "These mutations make these sodium channels almost—if not totally—nonfunctional." Even with pain stimuli, these sodium channels refuse to spark. The nerves neither whisper nor scream. They are silenced. "We are now in the unique situation of having shown gain and loss of function of SCN9A that results in either very bad or no pain. This is even stronger validation that the SCN9A gene is a master regulator of pain in humans." Waxman had proved the importance of this gene in our sense of pain by uncovering its role in two debilitating diseases.

Step by slow step, science is delineating and describing our pain pathways. The emerging pattern is unimaginably complex, with many connections missing but with the beginnings of coherence. We know now that the TRP proteins cluster at the tip of nerve endings, sensitizing them to pain and to temperature, whereas the SCN9A sodium channels are found all the way along the nerve's reach. "They are not, as far as we know, physically linked, but they work together and in the same way," Waxman told me. "The TRP proteins sense the external warmth and produce the first small depolarizing response in the nociceptor. I liken these to the antenna on a radio. The sodium channels next amplify this depolarizing response in the nociceptors and act like a volume knob so our brain gets to 'hear' our pain." The sand on a beach feels hot beneath our feet because our TRP proteins register its heat and alert the sodium channels. In a flash of neurochemical magic, they act together to spark action potentials that ricochet down the length of the neuron to the spine, then onward. On reaching the brain, the body's response causes us to perceive the sand's burning heat and leap into the cooling waves. If our feet have become a little singed, these nerves continue to fire, albeit more slowly, building the emotional feelings of pain that keep us in the cool water until we feel more comfortable.

*

In the summer of 2011, Waxman finally met the woman who had started him on his scientific quest. Pam Costa paid him a visit at Yale University School of Medicine. As Waxman walked her through the laboratory, they came across a picture on his computer screen of her mutated protein. Magnified to a resolution of a few angstroms, millionths the diameter of a human hair, the image showed unprecedented detail. The tangled clump of amino acids, zigzagging in three dimensions, appeared like a sculpture. She recalled, "I nearly lost my breath. For years I had desperately struggled to understand why I hurt, and here was the reason, displayed before my eyes. For the first time, I knew that my pain was not psychosomatic, that there was a real process underlining my daily suffering. It was the most validating experience of my life. It brought me and the researchers to tears." Finally, Waxman and his dedicated team are attempting to transform theory to therapy. They are trialing potential new medications in a handful of patients. "The early results are encouraging," he said. "It's an exhilarating time. We have a lot to do, but we are at last within reach of a cure for the rare people like Pam, and also for the many millions suffering with any form of neuropathic pain." His long-held dream may soon be a reality.

Aristotle excluded pleasure and pain from his list of senses, saying that both were "passions of the soul." This Aristotelian concept remains widely accepted. However, since Charles Sherrington, mainstream science has viewed pain as a sense in its own right, and today scientists wonder whether the same is true of pleasure. Given that pleasure and pain are wedded neurologically through the slow touch system, one might argue that they encompass one sense: the yin and the yang, the light and the dark of emotional touch. This sense reminds us how being sentient is as much about feeling as sensation. Aristotle may have proposed that the aim of the wise is not to secure pleasure but to avoid pain. Now we know that the wise would do well to listen to all the emotions in a touch. Pain may protect us from harm, but pleasure encourages us toward behaviors that secure our survival. The pleasure sensor may play an unexpected and more profound role.

Touch is not only the last of our senses to die but also the first to ignite when an embryo is only eight and a half weeks old. Francis McGlone argues that the warm liquid swirl of the womb, followed by a mother's attentive tendernesses, imprint on us the boundaries between our body and another's. "I think the early activation of this sensor is fundamental to the brain's healthy development," he told me. "It gives our brain the knowledge of you and me; it underpins and anchors our sense of self." Gentle touch seems to promote feelings of ownership, being in possession of our body. McGlone cites studies in which people even come to believe a lifelike rubber hand is their own when they watch it being stroked. The phenomenon is particularly powerful if the focus is on the upper, hairy side of the wrist. In addition, disturbances to a part of our brain that normally responds to gentle touch have been associated with converse bodily sensations. McGlone refers to patients prone to epileptic seizures reporting dramatic out-of-body experiences. Consequently, depending on the situation, gentle touch can foster sensations of either embodiment—even of fake limbs—or disembodiment. "I think these lines of evidence start to connect the hedonoceptor to our sense of identity."

If McGlone is right and the roots of our self lie in our skin, it follows there could be catastrophic consequences if their early natural activation is disrupted. McGlone points to distressing cases of maternal separation—in Romanian orphanages or among very premature babies—and ensuing incidences of psychiatric complications. He wonders whether faulty hedonoceptors might be a contributory factor in autism. "We know that people with autism lack empathy; their failure to understand another's intentions reflects their inability to understand self. We also know that many dislike gentle touch. Recently we found that the emotional parts of their brains fail to activate with gentle touch." His explanation for these observations is that their hedonoceptors are acting more like nociceptors: "I think the pleasure and pain they experience in touch have effectively become confused." McGlone is the first to concede that his theories require further research, but the early findings and their implications explain his demand for urgency. "I believe that touch is not a sentimental indulgence. It is a biological necessity in ways that we have yet to grasp. I am convinced this sensor will turn out to be the Higgs-Boson of the social brain: the missing nerve that socializes our developing brain."

The sense of touch has been as misunderstood as the vampire bat. Vladimir Nabokov wrote, "It is strange that this tactile sense, which is so infinitely less precious to men than sight, becomes at critical moments our main, if not only, handle to reality." As our world becomes more touch averse than at any other time in history, as the act of a touch becomes politicized and teachers are asked to refrain from close contact with children, as we lean toward conducting our relationships online and older people are said to be silently enduring an epidemic of loneliness, as we socially distance in the hope of quelling global pandemics, scientific evidence warns us to ignore this sense at our peril. It is not simply our handle on reality but the sense that, more than any other, makes us who we are.

6

The Goliath Catfish and
Our Sense of Taste

O N FEBRUARY 27, 1914, A LINE OF DUGOUT CANOES, LED by the Brazilian explorer Cândido Rondon and America's former president Theodore Roosevelt, disappeared into the Amazon's heart of darkness. The expedition would discover the Rio da Dúvida, the River of Doubt. "A river which was not merely unknown but unguessed at, no geographer having ever suspected its existence," claimed Roosevelt. "Whither it would go, whether it would turn one way or another, the length of its course, where it would come out . . . all these things were yet to be discovered." Nineteen men left, braving deadly snakes, rapids, and piranha-infested waters; only sixteen returned and, at one point, Roosevelt, delirious with malaria, begged to be left behind. Yet he lived to recount his adventures, including one encounter that became legend.

One evening, after well over a month on the river, two of the group returned from scouting downstream, hauling between them a strapping silver fish over a meter (3 feet) long. Roosevelt realized it was an unusual species of catfish with an "enormous head, out of all proportions to the body" and an "enormous mouth, out of all proportion to the head." Quite how unusual became apparent on gutting it, when they discovered the digested remains of a monkey. In *Through the Brazilian Wilderness*, Roosevelt wrote,

> Probably the monkey had been seized while drinking from the end
> of a branch; and once engulfed in that yawning cavern there was
> no escape. We Americans were astounded at the idea of a catfish
> making prey of a monkey; but our Brazilian friends told us . . . [it]
> occasionally makes prey of man.

He was also told that this reputed man-eater can reach 3 meters (10 feet) in length. Roosevelt had come face-to-face with the piraíba or the Goliath catfish, *Brachyplatystoma filamentosum*. His tales of this river monster have since lured many anglers to the shores of the Rio da

Dúvida, now also known as the Roosevelt River. Of all the predators in the Amazon, the Goliath is one of the largest and made formidable by the curious use of an often-overlooked sense.

Catfish are among the most ancient and diverse fish on our planet. Many of their three-thousand-odd species are found in the Amazon. The Goliath generally hunts other large fish in deep, fast-flowing tributaries, muddy and thick with fallen leaves from the overhanging canopy. As the foliage rots, it absorbs light, creating the river's famous blackwaters. Unable to rely on regular vision, the Amazon's fauna has evolved alternate ways to "see." The eye of the red-bellied piranha is said to register far-red-light wavelengths indecipherable to our eye, whereas the black ghost knifefish perceives its surroundings with electric fields.

All fish possess a lateral line along their flanks, highly sensitized to touch. Whereas we might feel on our skin the invisible wake of someone swimming nearby, fish can detect the tiniest turbulent eddies several meters away. The catfish also uses the four pairs of whiskers that inspired its name: two beneath its jaw trail along the riverbed, one pair protruding near its nostrils, and another oversized maxillary pair constantly on the move. These feel through the darkness, as might our hands, but touch is the least of their sensory talents. The Goliath wields its whiskers to track and close in on another fish meters away in the murk through the sense of taste. Research has revealed that of all vertebrates, catfish can detect the faintest trace of tastes. The longest recorded human tongue stretches just over 10 centimeters (almost 4 inches) from lip to tip. The giant anteater's can reach six times this length. Fully extended, the Malagasy giant chameleon's is not only even longer but also doubles its body size. However, catfish, and in particular the massive Goliath, take the organ of taste to new lengths. They do not possess a tongue as we know it. Their whole body *is* their tongue.

"The catfish is the iconic animal when it comes to taste. To describe it, I use the phrase 'swimming tongue,'" said Jelle Atema, recently retired from Woods Hole Oceanographic Institution on the Massachusetts coast. He is a sensory biologist with an affinity for aquatic creatures from mud snails to sharks, but his first love—and the focus of his PhD—was the catfish. "Lobsters taste with their feet, sea robins taste with their fins, but catfish taste with their entire body." Inspired by the American polymath Charles Judson Herrick, who almost a century before identified taste buds on the skin of catfish, Atema decided to take a closer look. His subject was the local bullhead, a catfish with much the same body shape as the Goliath, but at a fraction of its size, typically 25 centimeters (10 inches). He told me, "Although I had a modern microscope, I felt very much like a nineteenth-century scientist, in that my approach to begin with was simply to look. I dedicated hours upon hours to uninterrupted observation." First, he studied the bullhead's maxillary barbel. "Its leading edge was densely packed with mounds about 10 microns across," he said. "They looked like microscopic volcanoes rising up out of the skin, each with its own crater." Each volcano was a taste bud, each crater its pore. Using the more powerful scanning electron microscope, he saw the flask-shaped cells that register taste bunched together in groups. Their fine hairs erupted up through the pore and out of the crater. "The hair-like microvilli are where the taste cells come in touch with the outside world. There were so many that they looked like they were spewing out the volcano crater." Looking down the lens at these taste buds, marveling at their microscopic anatomy, he was struck by how similar they were to those in his own mouth. "The conservancy in taste buds between fish and humans is a reminder that we are the product of what originated in water many years ago," explained Atema. "Along with all other vertebrates, we inherited our taste system from our fish-like ancestors."

*

The middle of our tongue is scattered with red mushroom-like protrusions about a millimeter across. The folds of these aptly named fungiform papillae, as with others along the edges and back of our tongue, hold our taste buds. Like Atema's fish, each bud is made up of many taste cells buried in the tissue of our tongue whose microvilli reach up and out of its pore. We taste only what touches the tiny tendrils of these cells. Indeed, the word *taste* comes from the Middle English *tasten*, meaning to examine or test by touch. Foods do not themselves contain tastes. Just as our brain has evolved to interpret the different wavelengths of light reflecting off the mantis shrimp carapace as color, and the cresting and withdrawing of air molecules quivering the eardrum as the hoot of an owl, so too the sweetness of honey and the sharpness of sauerkraut are creations of our brain.

Our taste cells are primed with receptors, which are sensitive to salts, acids, sugars, and bitter compounds; when these are activated, we perceive salty, sour, sweet, and bitter tastes. More recently, scientists found glutamate or umami receptors, which lead to savory perceptions, extending our tongue's reach to five recognized taste channels. Even so, strictly speaking, much of what we think of as taste is not. "The tongue gets more credit than it deserves," declared the scientist and self-proclaimed smell chauvinist Avery Gilbert. "Taste is overrated." We do not taste chocolate but its sweetness. We do not taste lemons but their sourness. We do not taste vanilla at all. In a paper titled "What Aristotle Didn't Know about Flavor," Linda Bartoshuk and others pointed out that we all conflate taste and flavor. Avery observed that whereas English speakers use these two words interchangeably, other languages amalgamate them into one: *sabor* in Spanish, *Geschmack* in German, and *wèi* in Chinese. Yet most of a food's flavor—whether chocolate, lemons, indeed anything we chew—arises not from its tastes but from the airborne aromas it releases, which waft to the back of our mouth and up our nasal passage. Flavor reminds us that few sensory perceptions rely solely on one set of sensors operating in isolation. Our brain weaves different sensations into a coherent perceptual tapestry. Our experience of a landscape is most certainly seen, but it is also heard and maybe even felt or smelled as a breeze brushes skin and is caught on the tide of a breath. In the case of flavor, our brain hoodwinks us into thinking it is sensed by the tongue, when in fact most of the hard work is done by our nose. As any head cold shows, flavor is more about smell than taste. Yet despite taste's paucity, this sense is key to survival.

An experiment entailed swabbing the tongues of babies, only hours old, with different tastes. The results were immediate and dramatic. The newborns puckered their mouths and wrinkled their noses at the sour substance. They scrunched their faces in disgust at the bitter substance and spat it out. Then, when given the sugary solution, their faces lit up in an unmistakable smile. From the moment we arrive in the world, the sense of taste elicits strong emotions. The same love or hate of different tastes has even shaped language. In the throes of passion, Shakespeare's Juliet calls Romeo "my sweet," before telling him that "parting is such sweet sorrow." Teenagers call one another "salty" if upset or embarrassed, and a "sour look" expresses anger or resentment. Diane Ackerman added taste to her exploration of sensory metaphors, noting how "regret, an enemy, pain, disappointment, a nasty argument all are referred to as 'bitter.'" And a bitter pill threatens humiliation. So it is perhaps not surprising that a recent study showed how the simple act of having just read these taste metaphors on the page will have activated your brain's emotional centers.

Taste is as hardwired as our sense of pain and protects us in much the same way. Our distaste for bitterness and sourness warns us off poisonous or spoiled foods. Likewise, we are preprogrammed to derive pleasure from the first taste of our mother's milk and thereafter seek out foods brimming with energy-rich carbohydrates, whereas our enjoyment of salty and umami tastes leads us to the minerals and proteins that are necessary for life and growth. We have to find food tasty, as Ackerman observed: "It must tantalize us out of our natural torpor. It must decoy us out of bed in the morning and prompt us to put on constricting clothes, go to work." Smell might be learned throughout our lives, our preferences subject to the vagaries of culture, but our reflexive responses to taste rarely divide opinion. This holds true across mammals, birds, reptiles, amphibians, and fish, but the boundaries between the two senses can blur. "Catfish taste is so sensitive that they can taste food at a distance. They can detect chemicals a million times more dilute than we can, at concentrations not far removed from olfactory thresholds," Atema explained. "They use taste like we use smell." We may share structural similarities in our taste buds, but Atema discovered profound differences in endowment.

"I set about the perhaps dumb job of counting all the taste buds," Atema recalled wryly. Just as Ken Catania tallied touch organs on a star-nosed mole, Atema tallied taste buds on a bullhead catfish: not only on the animal's tip but also across its entire body, not only on one specimen but also on four more. "No one had done it before, perhaps for good reason, but I was young and curious." Peering down his microscope at the first specimen and using the millimeter-square measurement etched into its lens, he began, blissfully oblivious that these minute and meticulous observations would take months to perform. "I started with the whiskers and saw that the taste buds were so densely packed that I could not have squeezed more on had I wanted to." He counted over twenty-five taste buds in one square millimeter and some 20,000 on the four barbel pairs. "At this point it became obvious that these were not the tactile whiskers of a cat, but more like a set of eight external tongues." He found 3,000 on the lips alone, then moved to the body. The average back, belly, and sides, counted and recounted on all five fish, came to 155,000. "You can't touch the body of a catfish without coming into contact with vast numbers of taste buds." Next, he looked inside their mouths—along their gill arches, the sides, roofs, and bases of their oral cavities. The total taste bud tally on and in an adult bullhead was over 200,000. "The number still surprises people," he told me. "Although I've never counted them on a Goliath—and, for the record, would not want to—imagine how many more there could be on a fish at least eight times larger." Atema's discovery crowned the catfish nature's Taste King. Our tongues do not even approach these numbers. The *Encyclopaedia Britannica* claims we have at most 8,000 taste buds—a twenty-fifth the taste tackle of the bullhead, a far smaller fraction of the Goliath. However, an accident involving a glass bottle and an unforgiving laboratory floor would suggest that not all human tongues are created equal.

✳

The clumsy chemist was Arthur Fox, the lab belonged to DuPont, and although the chemical dust blew into the air almost a century ago, Linda Bartoshuk became intrigued by how people reacted. "Fox couldn't taste the phenylthiocarbamide powder, but the guy next to him thought it tasted awfully bitter," she explained. "They had the wit to run around and test their colleagues and found that a quarter of people, like Fox, could not taste it at all." In 2000, Bartoshuk was at Yale and about to become celebrated for debunking the human tongue taste map, thrilled to be simultaneously scoring a slam-dunk for her sex and her alma mater, telling me, "The idea our different tastes occupy different tongue areas is a myth perpetuated by none other than a Harvard man!" She turned her attention to the puzzling case from history. Fox was not the only case of taste blindness. She also knew of people who could be taste blind to another bitter chemical with yet another long name: 6-n-propylthiouracil, mercifully shortened to PROP. Furthermore, people who could taste both chemicals claimed to find other substances, such as caffeine, unusually intense. Such reports hinted at variations beyond simply tasting or not, but to ascertain whether they were significant, Bartoshuk needed an objective way to calibrate this sense.

"The question is, How do you measure sensation?" she asked. "You and I cannot share consciousness, so how do I know what you are tasting?" Scientists opt for magnitude scales, such as the ten-point experience scale. The previous chapter mentioned how pain is often measured this way. "But what if experiences differ?" she asked next. "In the hospital, we are medicated on the ten-point pain scale, but we know that childbirth means most women have endured worse pain than men, kidney stone sufferers excepted. So, a woman's 'four' is likely a more intense pain than that of a man and that means women are getting less medication for the same pain." Similarly, what if one person's "intensely bitter" is another's "mild"? How can we trust a magnitude scale to measure any taste? Such questions would reshape sensory research.

Bartoshuk's colleagues had shown that humans are remarkably skillful at comparing perceived intensities across the senses. "We are very good at matching the brightness of a light to the loudness of a tone," explained Bartoshuk, "so I began to wonder whether I could compare taste to a sensation not related to taste." For example, people were asked to match the bitterness of a chunk of dark chocolate to varying intensities of light. "Some matched the bitterness to the brightness of the moon; others to the low beam of headlights at night," she said. "We could then conclude that the second group perceived the same chocolate as more bitter." Next, the researchers asked the subjects to imagine a light scale in which a score of 100 stood for the brightest light ever seen, then again rate the chocolate's bitterness. On this sensory ruler, the first group rated bitterness at 14.9, the second at 32.4. "So now we had two measurements identifying two groups of people: one that tastes the same chocolate twice as bitter as the other." Soon the system revealed variations in tastes other than bitterness. "Put people in headphones," explained Bartoshuk, "and ask them to adjust the loudness of a tone to match the sweetness of a sip of cola." One group will set the tone to 80 decibels, about the loudness of a telephone dialing-tone purr; the other group will set it to 90 decibels, the loudness of a train whistle. "Given that 90 decibels sounds twice as loud as 80 decibels, I could conclude that the second group perceived twice as much sweetness in the cola as the first." She added, "We had been casting around in the dark but this cross-modal matching technique was finding variation in all our taste qualities. This was when we came up with the concept of the super-taster." Bartoshuk proposed that there are individuals within the wider population gifted with exaggerated taste sensitivities. The answer to whether such superhumans truly exist was, quite literally, waiting on the tip of her tongue.

"I have looked at more tongues than most people. Honestly, I can identify everyone in my laboratory simply by looking in their mouth," she told me. Working alongside the taste anatomist Inglis Miller, she set about researching this marvelously mobile sense organ. "Inglis taught us how to count taste buds." Accommodating volunteers would stick out their tongue to be swabbed with blue food coloring, inserted between two plastic cover slides held together by screws, then photographed through the lens of a microscope. Bartoshuk embarked on her own laborious bud-counting process. Whereas Atema had targeted individual taste buds in catfish, she opted for the structures that hold human taste buds. Also, rather than count every single fungiform papilla, she chose to sample them in the same area across different tongues. "A high school student had the brilliant idea of using the small hole left by a paper punch as our standard, so we set that on the photograph at the tongue tip, touching the midline, and counted." She saw vast variation between tongues, but a pattern emerged when she tested how their owners tasted PROP. "I have a pretty low count but my daughter has the lowest we ever recorded, only five papillae, and neither of us can taste PROP. We are the nontasters," explained Bartoshuk. "Meanwhile, the best super-taster to come into my lab, who tasted PROP as intolerably bitter, had a grand total of sixty-six in that tiny circle on their tongue." Bartoshuk and Miller had found the anatomical answer to supertaste, and it is a numbers game.

Most scientists now agree that we inhabit radically different perceptual worlds that are dictated by where we fall on the broad spectrum of taste-bud preponderance. Supertasters flaunt the highest number. They are the closest we humans come to the turbo-taste-charged Goliath. According to Bartoshuk, "Supertasters live in a neon taste world, where sweet, sour, salty, and bitter are sensed at least twice as intensely as regular tasters." Here again it seems that women have the edge over men. The star-nosed mole showed how increased sensor densities on typically smaller female fingers gave superior touch, but women are more likely to be supertasters thanks to larger numbers overall. That said, the neuroscientist Francis McGlone discovered a further intriguing correlation between our tongue's taste and touch abilities. Using tiles embossed with raised letters, subjects were asked to identify fonts of decreasing sizes using only the tips of their tongues. McGlone found differing touch proficiencies: "Supertasters were a quarter more tactually acute than regular tasters and twice as acute as nontasters." The two sensors are so entwined that higher numbers of taste buds naturally translate into greater densities of touch receptors, and thus supertasters also tend to be superfeelers.

Bartoshuk and her team have found that such people make up a quarter of the population. It may sound appealing to have such an embarrassment of sensory powers, but they are not natural gourmets. If anything, they find mealtimes fraught. "I've heard supertasters describe a dessert as being 'much too sweet.' As an extreme nontaster myself, I've never tasted anything in my life that was too sweet," she told me. "They particularly find foods too bitter. Leafy green vegetables can be a problem, Scotch whisky too, which to them is not just bitter but also produces a burning sensation." As the scientist Rob DeSalle put it, "Taste is a good case of 'more is *not* better.'" There is a silver lining: because supertasters find fatty foods cloying and distasteful, they tend to be slimmer, have healthier blood profiles, and have a lower risk of cardiovascular disease. Similarly, because they do not particularly like the taste of either alcohol or cigarettes, they are less likely to succumb to certain cancers. At the same time, supertasters are more vulnerable to a particularly peculiar disorder.

*

Dr. Raymond Fowler noticed something was awry when he was preparing dinner for his family. While the pasta was on the boil, he checked to see whether the cabbage was cooked by putting some in his mouth. It tasted surprisingly sharp. His daughter handed him a glass of cola to wash away the sourness. "It was like sulfuric acid," he said, "like the hottest thing you could imagine boring into your mouth." Soon the symptoms receded, but over the next few weeks, other odd tastes emerged out of nowhere. He sensed salt at the back of his throat, as if he were gargling with seawater. Yet his mouth was empty, and no matter what he did, the sensation would not fade. Eating became a game of Russian roulette: food would either lose its flavor entirely, reduced to "unsalted dough," or burn the sides of his throat as he swallowed. When even a glass of water became intolerably sweet, "as if someone had added three packages of Equal," Fowler could bear it no longer. He was, at the time, chief executive officer of the American Psychological Association, so he made an appointment with a colleague.

"When Ray walked into my office, I knew instantly that the distortions he described were what we call taste phantoms," explained Bartoshuk, who ran a taste clinic alongside her research laboratory at Yale. Just as people can be convinced by phantom limbs, they can experience nonexistent tastes. "The taste isn't there but feels real. It's not tied to one area but disembodied, as if coming from all over the mouth," she told me. "These phantoms are unpleasant, and they can be frightening. Doctors often fail to diagnose them, so people worry they signal that something is dreadfully wrong." Bartoshuk has a firsthand understanding of this fear. Her initial encounter with a taste phantom involved family tragedy and left such a deep impression that it motivated her to become a taste psychologist. "My father had been diagnosed with lung cancer, but at that early stage, his worst symptom was that food tasted strange. At the time, nobody understood what was wrong with him. Because of our research, I know now he was suffering from taste phantoms." Phantoms tend to be salty, sour, sweet, or bitter, but they can also be metallic and even create burning sensations. Fowler was experiencing the full gamut. Bartoshuk had learned through trial and error that anesthetizing the taste buds of the tongue would not rid him of them. Far from it. "The first time I did that to someone—a lady with a salty phantom—her phantom in fact got much worse, much saltier. I'll never forget it. It was awful. Everyone was stunned but that was when we realized that these phantoms don't start on the tongue but in the brain." The phantoms besieging Fowler and others remind us again that taste buds alone do not create taste. To gain a better understanding of why some individuals come to believe in such ghosts, we turn once more to the catfish.

*

While Jelle Atema was focused on the bullhead catfish from the outside in, another biologist, at the Massachusetts Institute of Technology, approached it from the inside out. Tom Finger, now codirector of the Rocky Mountain Taste and Smell Center, was just starting out as a graduate student. Having prepared paper-thin sections of the catfish's postage-stamp-sized brain, he looked at them under a microscope.

He could just discern shapes, which seemed like anatomical divisions. "Once you know what you're looking for, the subtle shading differences on the surface of the catfish brainstem medulla reveal its entire body," he told me. Finger saw a miniature silhouette of the fish, from barbels to tail fin. "I called this little fish *piscunculus*." Just as the brain maps of the star-nosed moleunculus and our homunculus represent their respective bearers with body dysmorphia, Finger found that the piscunculus recasts the catfish as having a small body, outsized by gargantuan whiskers. "Its maxillary barbels accounted for the same space as the rest of its body." The piscunculus was a touch map like the others but also the first known *taste* map. Finger had exposed the cerebral dimension to Atema's surface taste buds. Where Atema saw high densities on the body, Finger saw large areas of the brain dedicated to their input. According to Finger, "The amount of brain space given to the whiskers reveals that they are the catfish's taste fovea, just as the mole's star and a human's fingers are their touch foveas." The piscunculus processes information from the taste buds on the outside of the catfish's whiskers, trunk, and tail but, unlike the moleunculus and our homunculus, also from those on the inside of its mouth. Finger began to follow the delicate tracery of nerves connecting the bullhead's brain to its taste buds. Meanwhile, Atema worked in reverse: "Regular forceps were too big to grab individual nerves. I needed them so fine, about 50 microns across, that I had to engineer them myself." Together they isolated three principal nerves that form two sensory pathways. This begged the question whether catfish might have more than one sense of taste.

We know now that as a streamlined Goliath glides through the soupy Amazonian blackwaters, it relies not only on its body's lateral line but also on its taste buds to perceive its surroundings. "Fish and other animals are leaky bags whose skin and gills slowly release various chemicals, including amino acids," said Atema. Catfish taste buds, primed to amino acids, are among the most sensitive in the animal kingdom. They detect dilutions as low as one part amino acid to 1 billion parts water. A piranha cuts through the current, just ahead and to the side. Amino acids in its wake reach taste buds on the Goliath's maxillary whiskers at slightly different times. In much the same way as the great gray owl compares sounds arriving at each of its two ears, the fish compares tastes arriving at these whiskers to steer its course. "Catfish use this barbel taste fovea to explore a taste plume in three dimensions; in fact, they use their whole body as a sensory plane to track prey," added Atema. When the Goliath is within striking distance, a flick of its tail launches its jaws at the target.

Atema's and Finger's anatomical studies proved that the body taste buds are served by the seventh cranial nerve, one of the only three nerves that carry taste fibers. Running from the body to the brain, it is the sole conduit of taste information from the outside world. Atema explained, "The seventh cranial nerve innervates the external taste sense that makes the fish alert to the presence of food and guides it straight to the source." The two other taste nerves—the ninth and tenth cranials—provide a different sensory route. They wire the taste buds of the fish's mouth and gill arches to its brain. Their input has another purpose and will kick in only when the Goliath's jaws wrap around the piranha. "Catfish have two separate senses of taste," Atema said. "As well as the external exploratory sense, they have an internal sense that controls whether the fish will swallow what is in its mouth." Unfortunately for the piranha, the odds are not in its favor. The separation of taste into two senses is not unique to catfish. Lobsters and insects also have internal and external tastes. Humans may miss out on being able to taste through their feet or fingertips, but maybe our tongue conceals hidden taste talents.

Taste information is carried from our tongue to our brain by the same three nerves found in the Goliath. The fungiform papillae on the tip and first two-thirds of our tongue are innervated by the seventh cranial nerve and its offshoot, the chorda tympani. The papillae at the rear of our tongue and larynx are innervated by the ninth and tenth cranial nerves. This division has repercussions for how we taste. Finger argues that we use the front of our tongue, much like the catfish uses its body taste, to explore the world. "You might lick something before putting it in your mouth, especially something you haven't encountered before," he told me. "Just watch how babies put things in their mouth and use their tongue to examine them." We rely on the back of our tongue in much the same way as the catfish uses its internal taste. "Once the potential food item reaches the back of the tongue, you're making a decision: Should I swallow this, or not? At this point taste also primes the digestive system for the food it is about to receive." Otherwise, Atema added, "it triggers the spit, rather than the swallow, response and perhaps even the gag reflex."

Our taste, like that of the catfish, bridges the external and internal worlds. Consequently, one can argue that our tongue harbors two discrete senses of taste. Finger said, "Taste at the front of the tongue is more an exteroceptive sense, and at the back it is an interoceptive sense." Atema elaborated: "The division is not as clear-cut as in the catfish, but just bringing up the question of whether we should consider our exteroceptive and interoceptive tastes as separate senses is already an important step in understanding just how different these two systems are to one another." If it were not for comparisons with the catfish, then an appreciation of our tongue's subtle multiplicity might well have escaped us. There is a practical benefit to understanding the nerves of our two taste pathways. The way in which they interact holds the power to exorcise taste phantoms.

*

Linda Bartoshuk's research was also homing in on the chorda tympani branch of the seventh cranial nerve. Cases with damage to this particular nerve kept presenting in her clinic. "The chorda tympani is actually easy to injure," she said. "It runs from the tongue through the middle ear, so it can be affected by ear and upper respiratory tract infections." Despite this, the patients rarely complained that their taste was impaired. Bartoshuk decided to test what would happen if she anesthetized this nerve in healthy individuals. She found that their whole mouth taste was unaffected, which was consistent with what she had seen in her patients. She theorized that as soon as the seventh cranial nerve offshoot fails, whether through damage or anesthesia, the ninth and tenth take over. "The three cranial nerves carrying taste inhibit one another in the brain. If one is damaged, the release of inhibition on the others means they compensate for the loss of input." The two sensory pathways behave as if in conversation; as soon as one falls silent, the other increases the chatter. This ensures that the brain continues to create a full taste perception. The study's arguably more significant discovery happened once the experiment was over and the subjects had left for lunch.

Bartoshuk knew something was afoot from the first volunteer: "As she chewed on a bagel, watching the sides of her mouth which were still numb from the anesthetic, she told me that a strange salty sensation grew at the back of her mouth." Eight of the sixteen subjects reported similar experiences. Two had phantoms that were salty, one sweet, one bitter, one salty sour, one sour metallic, another bitter metallic, and the eighth had a bitter phantom that changed to sour, then sweet. It was a full haunted house. "They were startling, serendipitous even," she recalled, "but they showed that taste phantoms are actually a reflection of the way the taste system is wired in the brain." Bartoshuk believes that the conversation between the three taste nerves has an evolutionary purpose. Along with our basic ability to taste, it serves survival by safeguarding the sense in animals with nerve damage, but sometimes the volume is cranked up so high that it creates perceptions that do not exist. "The release of inhibition to the brain is so energetic that I think what happens is you may actually be tasting 'noise' in your nervous system." Ultimately this insight would show the way to a cure.

"We now have a drug that simulates inhibition in the brain," Bartoshuk informed me. If the chorda tympani is injured, this medication can redress the balance of nerve chatter and quiet the phantoms. "If I could go back in time, knowing what I know now, with the drugs that we have now, I feel certain I could have helped my father and made his last days more bearable." Meanwhile, Fowler's phantoms disappeared without the need for intervention; Bartoshuk suspected his chorda tympani had only been temporarily incapacitated by a viral infection. Finger said, "Linda's discoveries of how the brain processes sensation and creates phantoms are fascinating. However, the truth is that despite substantial advances in the last few years, we still don't know much about how taste is processed cortically. Our understanding of how taste perceptions arise is far behind that of vision, hearing, touch, and, for that matter, smell. This is a big debate in the field right now." As our knowledge expands, our definition of this sense, or senses, shifts. A discovery made in 2014 adds an unexpected twist to the tale of the catfish.

✳

Marine bristle worms are masters of concealment. They live beneath the ocean bed in tunnels, disturbing the surrounding water only when they emerge to feed. The overlying sediment masks them as well as their scent. Yet even on the blackest nights, the Japanese sea catfish shows swift precision in locating their burrow entrances, encircling them with its lips, and sucking out any unsuspecting occupants. John Caprio, a neurobiologist at Louisiana State University, became intrigued by how the catfish senses these worms. He is an expert in using electrophysiology to record the activity of fish nerves. Much of his day is spent sequestered in a Faraday cage, a contraption made of metal wire mesh that shuts out any ambient voltage. "The minute action potentials that I record would be drowned out by the voltage in all the wires running through my laboratory building," Caprio told me. With the fish sedated, he hooks and probes their thread-like nerves with ultrafine electrodes to intercept their signals. "I see them fire on an oscilloscope, but I also hear them. The system is set up so a nerve finding is heard like the rattle of a machine gun." His prior research on catfish taste nerves revealed how uniquely sensitive their external taste buds are to certain amino acids, enabling the detection of prey many meters away. He suspected that taste might have a role to play in this feat of worm charming.

"Science can be about serendipity; that's why I love it," Caprio admitted. "You have to keep your eyes and mind open to possibility." With an electrode poised to pick up any excitation in the taste nerves, he filtered various amino acids over the catfish's whiskers. Then something odd happened. One group of amino acids triggered a signal so powerful that it stood out clearly from typical taste responses. In search of an answer, he turned to his textbooks and found a table displaying widely differing alkalinities and acidities of amino acids. "That's when it hit me," said Caprio. "I realized these amino acids were subtly changing the pH of the solution that I was filtering across the barbel. That's what the Japanese sea catfish was sensing." When bristle worms exhale, they release carbon dioxide that incrementally increases the acidity of the otherwise mildly alkaline seawater. The catfish's whisker is so sensitive to such tiny and transient bursts of acidity that the worms do not stand a chance; they can hide, but they cannot hold their breath. "It's ingenious, astonishing," he said. "It's the first time that a sense for pH has been described in any creature. Although I found the response in the taste nerves, it might not be taste. And if it is taste, it is not as we know it." Caprio wonders whether his discovery might be part of a wider revolution underway in the field—one that threatens to rewrite what we mean when we talk about taste.

Solitary chemosensory cells are a hot topic in taste research. Although they look like taste cells, they are found singly rather than gathered into taste buds, as their name implies. They were first discovered scattered on the skins of fish. "Catfish, hagfish, lampreys, sharks, bony fishes like trout, zebrafish, minnows." Tom Finger reeled off a list, before adding, "Frankly every fish so far examined." Scientists have since spotted similar cells in other vertebrates: amphibians, reptiles, mammals, and, most recently, humans. Our tongue is not the only repository of "taste" cells in our body. Finger and others have identified taste-like cells, called tuft cells, lining our airways. "We see them in our respiratory system, in our noses and our windpipes," he told me. "Similar taste-like cells also occur along our digestive system from the stomach, the gallbladder, bile duct to the small intestine." It seems the more we look within ourselves, the more we find. Unlike the taste cells in our tongue's papillae, the scattered taste-like cells never lead to any conscious bitter, salty, sweet, sour, or savory perceptions; they "taste," but we do not. They work beneath our awareness. "These cells show that taste isn't just for taste buds. The same molecule that in our mouth creates a bitter sensation will do something different in the airway and gut," explained Finger. "For example, in the nose, the same bitter-tasting molecule creates an itch so we sneeze." These bitter receptors do not create taste sensations, but prompt our body to eject airborne toxins or pathogens forcibly. They also instruct our immune system to marshal a response in case any unwanted visitors sneak through.

The taste cells in our digestive system serve yet another purpose. The sweet and glutamate receptors in our stomach release an appetite promoter, perhaps explaining why that first mouthful of aromatic stew or sweet soufflé leaves us wanting more. Finger explained, "In much the same way that our conscious sense of taste is both a guardian and a guide when we eat food, these solitary taste cells guard and guide us, but in very different ways and mostly unconsciously." Their discovery raises fundamental questions. "These are taste receptors, but is this taste?" Finger asked me. Referring to Caprio's work, he added, "They may account for the pH sense that John has seen in the Japanese sea catfish, but we don't yet know. No one has defined what solitary chemosensory cells are in fish, let alone in humans. We don't know what these cells are doing. We don't even know what they mean." Their mere existence shakes the foundations on which our understanding of this sense stands.

*

Our tongue can talk, tease, and taunt, but it can also elevate us to the realms of supertasters. The Goliath catfish and its many cousins show that our sense of taste is more diverse than we could have imagined, and its reach extends beyond that of our tongue. "It is hard to get a handle on what it means to be human," said Jelle Atema. "We must look to the creatures with whom we share our planet to inform who we are. It is far easier to get an idea of how taste works in a catfish or how smell works in a lobster"—or, for that matter, how sight works in a spookfish, hearing in a great gray owl, or touch in a vampire bat. Through these animals, our knowledge extends back millions of years. "There are many parallels to the way taste at its simplest, most fundamental level is processed in fish and in humans because we evolved from a fish-like ancestor," said Tom Finger. Yet even before this, further back in time, when life on earth was starting, scientists think the creature that first tasted the world did so with solitary chemosensory cells. Over the ages, evolution honed this unconscious sense into the sophisticated taste system seen today in fish and humans. Solitary chemosensory cells may be the origin of every single taste taken since, every mouthful rejected or relished. One hundred years ago, when Theodore Roosevelt stepped into his canoe, oar in hand, and braced his shoulders and chest for that first stroke into the unknown, he could not know what lay ahead or the tales he would bring home of the deadly jararaca pit viper, the "swarms" of red-bellied piranha, and the man-eating Goliath. Today science too sets forth into terra incognita, hoping to elucidate and redefine our much-neglected sense of taste.

7

The Bloodhound and
Our Sense of Smell

A PATIENT OF OLIVER SACKS RECOUNTED A REAL-LIFE Kafkaesque metamorphosis. One morning, "Stephen D., aged twenty-two, medical student," found himself altered. "I had dreamt I was a dog," he told Sacks, "and I now awoke to an infinitely redolent world—a world in which all other sensations, enhanced as they were, paled before smell." He recalled visiting a perfumery: "I had never had much of a nose for smells before, but now I distinguished each one instantly—and I found each one unique, evocative, a whole world." He started identifying friends and colleagues by smell: "Each had his own olfactory physiognomy, a smell-face, far more vivid . . . than any sight face." Stephen D. used his keen canine nose to find his way around New York City. He was compelled to sniff everything, feeling that otherwise "it wasn't really real." He likened his newfound perception to having been color blind, then finding himself in a world of color. Three weeks later, the strange transformation ceased abruptly, and Stephen D.'s sense of smell returned to normal. Subsequently, Sacks would learn that his patient had been using cocaine, PCP, and amphetamines, so he diagnosed "an amphetamine-induced dopaminergic excitation," resulting in an unusual case of hyperosmia, from the Greek *hyper* and *osme*, meaning excess odor. He wrote up the case notes under the title "The Dog Beneath the Skin," with his patient concluding wistfully, "It was like a visit to another world, a world of pure perception, rich, alive, self-sufficient, and full. If only I could go back sometimes and be a dog again!" Stephen D. had the entire animal kingdom at his disposal, yet he had fixated on one creature. His choice reveals a greater truth; when it comes to olfactory prowess, few surpass the humble hound.

As director of the Sensory Research Institute at Florida State University, James Walker conducted two studies that pitted the human nose against that of "man's best friend." In 2003, using methods that enabled him to test trace samples of amyl acetate—the odor of ripe banana—Walker looked at humans, then, three years later, at dogs. The canine thresholds—the smallest amounts smelled—were 20,000 to 30,000 times lower than previous reports; dogs were able to detect dilutions of one to two parts per *trillion*, the equivalent of sensing a drop in an Olympic-sized swimming pool. When compared to humans, Walker wrote, "Our recent investigation of human odor detectability yielded thresholds approximately 10,000 to 100,000-fold higher than those we report here for the dog." The studies are often cited as proof that a dog's nose is up to 100,000 times more sensitive than ours. Little wonder that trained cadaver dogs can sniff out drowned bodies beneath fast-flowing rivers and trained avalanche dogs can find people buried by as much as 7.2 meters (24 feet) of snow. Recently a team from the International Forensic Research Institute in Florida fitted cars with pipe bombs, which they detonated to prove that bloodhounds detect someone's scent on the bomb after it has been blown to smithereens. Of all dogs, this breed has long been celebrated for its nose. There are bloodhounds carved into a stone slab at the Assyrian royal palace in Nineveh, dating from the seventh century BCE, but one in particular is hailed as the king among kings.

"No book on bloodhounds could be complete without mention of the greatest of them all, Nick Carter," wrote Catherine Brey and Lena Reed in *The New Complete Bloodhound*. "The dog was the greatest who ever lived," declared Leon F. Whitney in *Bloodhounds and How to Train Them*. Born in Lexington, Kentucky, in 1900 and named after the famous detective of American dime novels, Nick Carter was the best sleuth in hound history. He trailed for Captain Volney G. Mullikin, and, according to Whitney, the duo became so celebrated that when they "came to town to trail, great crowds gathered." Together they put over 600 criminals behind bars: 126 in one year alone, with what Brey and Reed called "a 78% conviction rate, a record many a human detective would envy." In one noteworthy case, Mullikin and Carter had been called to the scene of an arson attack: a henhouse had been burned to the ground. They arrived more than a hundred hours after the match had been struck. The trail had been subject to the vagaries of wind and weather for over four days and nights. Despite this, Carter managed to find, then follow, the arsonist's scent to a neighboring village and to one front door. Mullikin knocked. When a man appeared, he asked, "You weren't counting on the dog when you burned that hen-house, were you?" The man's simple "No" was a tacit admission of guilt. The dog's nose had been so severely tested that Whitney concluded, "Nick Carter holds the world's record for a one hundred and five hour trail that resulted in a confession."

Science has proven the impossibility of leaving without a trace. Humans have been shown to shed a million cornflake-shaped, aerodynamic skin cells every minute; there are up to 50,000 in every liter of air we inhale, and 500 settle in every meter we walk. The hound's pendulous ears helped sweep these scent molecules toward its muzzle. Its folds of wrinkled flesh, the shawl, and drool hanging from its lips may well have caught others, but the bloodhound's excellence rests in its snout.

*

Gary Settles was forced to switch his research away from the fluid dynamics of high-speed airplanes. "Funding dried up in the decade after the Cold War, so I was actively seeking new problems to solve," he told me. The US Defense Advanced Research Projects Agency requested that he turn his attention to the dog. "The dog's nose is universally recognized as the gold standard of olfactory acuity, yet I was shocked to discover that no one had ever looked at the fluid dynamics of how it smells, how odor molecules enter its nose." Settles, who runs the Gas Dynamics Laboratory at Pennsylvania State University, decided on a highly specialized form of image capture called Schlieren photography. Invented in 1864 by a German physicist, it uses smoke and mirrors to, quite literally, make invisible air currents visible as they flow around objects. Settles found a willing subject in a colleague's pet golden retriever. "We were so lucky with Bailey. She was a beauty and such a well-behaved dog. Also she would do almost anything for some Campbell's beef broth." All breaths come in pairs. The research team was about to discover how different they are in a dog.

"We trained Bailey to rest her head on a wooden block, keeping still so we could film her sniffing in slow motion." They recorded her in-breath at 1,000 frames per second, placing a cup of her favorite soup at increasing distances from her nose. Played back, the Schlieren footage showed Bailey's inhalation drawing a thick plume of odorant molecules from the soup when it was 10 centimeters (4 inches) away. "That was an unexpected reach," Settles told me, but Bailey's out-breath held more surprises. The dog's comma-shaped nostrils lie beneath the alar fold, a flap of skin. Studying the footage, the team could see this flex between breaths; it opened on the inhalation, to allow air in through the wide-open nostrils, then closed on the exhalation, so that the air was forced through the tail of the comma, diverted sideways and away from the food that Bailey was sniffing. "It means any expired air doesn't travel back against the incoming air flowing from a scent source," Settles explained.

Next, computational analysis showed how this constricted nostril shape extrudes air as it leaves the dog's nose to form two co-rotating vortices that spin and create suction. "These expired air jets entrain the surrounding air, increasing the draw of an air current toward the nostrils." This extends the nose's reach even farther and enables it to gather scents from concealed locations, whether from under stones or inside crevices. "We were seeing how canine nostrils worked for the first time," recalled Settles, "and we realized they are more than just simple orifices; they are unparalleled odor-capture devices." The dog's flexi-directional and jet-assisted sniff is, like the ruffed face of the great gray owl, finely adapted to gather as many sensory cues as possible. To visualize where the air goes next, the researchers created the first three-dimensional detailed image of the inside of a dog's nose.

Smell starts in dogs and humans when scent molecules are caught on an incoming tide of a breath, wash through to the depths of our nasal cavities, and reach the olfactory epithelium. This moist and mucous-coated membrane is crammed with olfactory neurons—the nose's answer to the cones and rods of the eye, the hair cells of the ear, the mechanoreceptors or touch neurons of the skin, and the taste buds of the tongue. The business end of these long cells bears microscopic cilia that stick out and wave in the air current. These hairs are coated with olfactory receptor proteins that snare the passing scent molecules. The American neuroscientists Linda Buck and Richard Axel made the unanticipated and astonishing discovery—for which they were awarded the first-ever Nobel Prize for olfaction in 2004—that mammals share an enormous family of genes that code for these receptor proteins. So the noses of dogs and humans are primed with the same kinds of scent-snagging biology. Furthermore, the Axel and Buck collaboration showed that each olfactory neuron bears only one receptor, which interacts with specific scent molecules in much the same way as a lock opens only to the right key. Aromas are made up of many odor molecules, so having more receptors increases the range that can be detected. Dogs lay claim to some 800 functioning receptors, therefore 800 types of olfactory neurons, whereas we have half that number. In addition, studies suggest our olfactory tissues are not as expansive as those found in dogs. Alexandra Horowitz, in *Being a Dog*, notes wryly that if the "olfactory epithelium were spread out along the outer surface of the dog's body, it would completely cover it," whereas in humans, it might "cover a mole on our left shoulder." Whatever the precise quantities, it is widely accepted that the olfactory epithelia of bloodhounds have hundreds of millions more olfactory neurons and twice the types, promising both greater smell sensitivity and range than humans. However, the epithelium's position in the snout puzzled Gary Settles. In humans, it is located a finger's length up a nostril, but in dogs, it is at the very end of their prominent muzzle and tucked away in a recess. "Set aside like a cul-de-sac," said Settles. "We hoped to answer how that could possibly work."

An MRI scan of Bailey's head revealed the intricately folded, fine scroll-like bones that make up the recess. "We could see these ethno-turbinates in unprecedented detail," Settles told me, "and it reminded us of the immensity of the dog's olfactory surface area." They could also see how it was separated from the main airflow by a horizontal plate of bone. Combining data that had been collected on the volume and speed of Bailey's sniff with the results of the scan, they created a computer simulation of how air currents whirl in, around, and out of a dog's nose. For the first time, they witnessed how an odor molecule, caught by a nostril, passes through the nasal vestibule; on reaching the bony plate, it either gets sucked down the breathing route into the lungs, never to be smelled, or takes the high-speed vertical ride up toward the olfactory recess. "When we saw how the fast and chaotic flow in the vestibule instantly slows down on reaching the larger space of the olfactory recess, we realized that the cul-de-sac is key." Humans do not have a recess, so air whistles through our nose at 27 liters (6 gallons) per minute, exposing our olfactory epithelium to gale-force winds. The dog's recess enables more odor molecules to settle on the cilia, then find and unlock their specific receptor proteins. "I could never have guessed what Bailey would teach us," Settles said. "A fluid engineer would struggle to come up with a better solution. Evolution created the best scent-capture device I know of with the dog's nose." Yet catching scents is only the start—alluring fragrances or rank odors have to be smelled—and just as color vision can be better understood through its complete loss, so too can smell.

*

Some time after Oliver Sacks met the student with hyperosmia, he came across his antithesis. Rather than an excess of hyperosmia, the man had anosmia: the sense had deserted him entirely. "Sense of smell?" the man queried. "I never gave it a thought. You don't normally give it a thought. But when I lost it—it was like being struck blind. Life lost a good deal of its savor. You *smell* people, you *smell* books, you *smell* the city, you *smell* the spring." He mourned all that he had taken for granted, adding poignantly, "My whole world was suddenly radically poorer." The impacts of such a condition are known to be both insidious and manifold. As flavor is mainly smelled, food dwindles to our five basic tastes, robbing mealtimes of their pleasure. Some people lose their appetite and lose weight; others eat to feel satisfied so gain weight. Many become haunted by notions that they can no longer detect their own body odor, spoiled food, or fires. Precious memories that were tethered to certain fragrances are lost. Indeed, anosmia can give rise to depression, and its effects are frequently described in devastating terms. One person said, "I feel empty, in a sort of limbo"; another, "It is almost as if we have forgotten how to breathe." One woman complained of feeling so disconnected that her personality changed; her husband said he no longer recognized her.

Sadly, a loss of smell is not uncommon. Like the loss of taste, it can be caused by viral infections. In 2020, anosmia became familiar to many as an unpleasant quirk of COVID-19. But these symptoms can pass. When Sacks next saw his patient, this seemed to be the case. "To his astonishment and joy, his favorite morning coffee, which had become 'insipid,' started to regain its savor. Tentatively he tried his pipe, not touched for months, and here too he caught a hint of the rich aroma he loved." Yet on being reexamined, the patient was told that nothing had changed physically: "There's not a trace of recovery. You still have a total anosmia. Curious though that you can still 'smell' your pipe and coffee."

Sacks learned that the patient's anosmia had begun after he had sustained a head injury that had damaged the olfactory tracts of his brain. Although his nose was still inhaling aroma-laden air and his olfactory receptors were still hooking odor molecules, his brain scrambled their signals from becoming sensation. Sacks suspected these new smells were the result of a "controlled hallucinosis, so that in drinking his coffee, or lighting his pipe . . . he is now able to evoke or re-evoke these, unconsciously, and with such intensity as to think, at first, that they were 'real.'" Arguably, such scent chimeras are not dissimilar to Stephen D.'s possible drug-induced sensory metamorphosis. In fact, olfactory hallucinations take various forms, few of which are appealing. Parosmia describes the condition when certain odors give rise to distorted perceptions; one sufferer awoke to find everything smelled of burned toast, another to gasoline. Phantosmia is smell's equivalent to taste phantoms, in which odors are entirely imagined. Without doubt, the worst is cacosmia, where the stink of sewage becomes inescapable. All these perceptual distortions remind us that smell exists within us.

✳

Recall the questions posed in the first chapter: "If a tree falls in the forest, and no one is around to hear it, does it make a sound?" And, "If no human eye is around to view it, is an apple really red?" In *What the Nose Knows*, the olfactory scientist Avery Gilbert puts a match to the fallen tree: "A tree burning in the forest does not smell if no one is there to smell it." Smell is not present in the molecules that stimulate the receptors of our olfactory neurons. As the neurobiologist Gordon Shepherd observed, "When smell molecules bounce in and out of a receptor-binding pocket in an olfactory receptor neuron, all an individual cell 'knows' is how much the features of the smell molecule have tickled its binding sites. The greater the tickle, the more the cell responds by generating impulses." The brain reads the differences between the responses of different olfactory neurons to create perceptions of smells; precisely how the brain does this remains one of its best-kept secrets. Such understanding as we have began in the nineteenth century and involved a Frenchman with a penchant for human anatomical dissection.

During the course of his life, the surgeon and anatomist Paul Broca amassed 432 human brains, preserving them in jars of formalin. On his death in 1880, and as he had requested, Broca's own brain joined the collection. A century later, the American cosmologist Carl Sagan made a pilgrimage to the Musée de l'Homme in Paris to see it and wondered, "How much of that man known as Paul Broca can be found in that jar?" According to Oliver Sacks, Broca "opened the way to a cerebral neurology, which made it possible, over the decades, to 'map' the human brain, ascribing specific powers—linguistic, intellectual, perceptual, etc.—to equally specific 'centres.'" Today, Broca is acclaimed as the founder of modern neuropsychology. He was particularly interested in a region of our forebrain known as the olfactory bulb, which lies just behind the olfactory epithelium. A closer look reveals their intimate connection: the long olfactory neurons of the epithelium actually reach through the bony and perforated cribriform plate, directly into the olfactory bulb. One nerve is all that separates a scent that has wafted up our nose and landed on our olfactory epithelium from our olfactory bulb; one more and you reach the olfactory cortex. As the neuroscientist Stuart Firestein exclaimed, "You can get from the outside world to cortical tissue in the brain in two synapses—*two synapses*! In the visual system, you'd still be in the outer retina." No other sense accesses the brain's core cortex in so few steps. Perception starts as an odor's various molecules unlock specific receptors and their neurons fire into the olfactory bulb.

Relative to the overall size of the brain, Broca noticed that our olfactory bulb is proportionately smaller than those of other mammals. He saw the reverse pattern with another part of the brain: our frontal lobes are proportionately bigger. The year before his death, Broca wrote in the *Revue d'Anthropologie*, "This lobe, enlarged at the expense of the others, grabbed the cerebral hegemony; the intellectual life is centralized there; it is no longer the sense of smell that guides the animal: it is simply intelligence enlightened by all the senses." He believed that during evolution, the olfactory bulb had made way for the frontal lobes and our sense of smell had been traded in exchange for free will. "This," he claimed, "is the cause of the atrophy of the olfactory apparatus." To Broca, humans were not simply inferior smellers; they were "nonsmellers." His conclusions became the standard view of the twentieth century. Sigmund Freud threw his weight behind the idea, arguing that smell was of interest only to children, perverts, and neurotics. He stated, "The organic sublimation of the sense of smell is a factor of civilisation." Its loss supposedly raised us above other animals, effectively civilizing and humanizing us. Our species was relegated to the status of microsmats: "tiny smellers" to the dog's macrosmatic supremacy. In G. K. Chesterton's famous poem about a dog, "The Song of Quoodle," Quoodle laments on our behalf, "They haven't got no noses, / The fallen sons of Eve; . . . // The brilliant smell of water, / The brave smell of a stone." However, contrary to the textbook wisdom of Broca and Freud, evidence is amassing that calls into question the notion of human microsmaty.

✳

The deaf-blind writer Helen Keller is often held up as a scent prodigy. She called smell the "fallen angel," and along with the sense of touch, it was how she apprehended the world. She described how she could distinguish men from women, babies from adults, one person from another, simply from the way they smell. "I have not, indeed, the all-knowing scent of the hound," she admitted:

> Nevertheless, human odors are as varied and capable of recognition as hands and faces. The dear odors of those I love are so definite, so unmistakable, that nothing can quite obliterate them. If many years should elapse before I saw an intimate friend again, I think I should recognize his odor . . . as promptly as would my brother that barks.

Despite popular belief, scientific studies have not shown that blind people have a superior sense of smell. Quoting six scientific studies over the past two decades, Avery Gilbert observed that "without exception, they find that the blind are no more sensitive than the sighted—both groups detect odors at about the same concentration. Nor do blind and sighted people differ in the ability to discriminate one odor from another." If people who are blind possess an advantage—and only three of these six studies suggest they do—it is simply that they are better at naming odors, so perhaps Keller's gift was not so miraculous after all.

In *"Surely You're Joking, Mr. Feynman!"* the theoretical physicist Richard Feynman recounted the time he decided to explore his inner bloodhound, enlisting the help of his wife. "Those books you haven't looked at for a while, right?" he said, pointing to a line of books on the shelf. "When I go out, take one book off the shelf, and just open it—that's all—and close it again; then put it back." She duly did as requested; he returned and pressed his nose up against each book in turn. He then correctly indicated the book that she had handled. He recalled that there was "nothing *to* it! It was easy. You just smell the books. It's hard to explain [but] you can tell. . . . Humans are not as incapable as they think they are: it's just that they carry their nose so high off the ground!" Soon it became his party piece, and he would open with the gambit, "Because I'm a bloodhound . . ." Perhaps it amused him to note how onlookers were more inclined to assume they had been duped than that he could smell the subtle traces left by people's hands. Feynman may well have been an intellectual giant, but his eyes and nose were no different from yours or mine. In the past decade, there has been a flurry of scientific studies suggesting that we all have Keller's gift and any one of us could astound audiences with Feynman's party trick.

*

Matthias Laska, a sensory physiologist at Linköping University in Sweden, is a world authority on the smell sensitivity of muzzles, bugles, and beaks across the animal kingdom. "My wife says that if I have one talent," he confessed, "it is that I can think like an animal." Humans have been tested with over 3,000 individual scents, but only 17 of the 5,500-odd mammal species have been studied, with only a meager 138 odors. Laska has spent the past three decades trying to redress the balance. "Comparing smell sensitivities between species is not without its problems," he said. "Even if it was possible to use the same method with different species, this would very likely put one at an advantage over another." Consequently, much of his research has been to invent new ways of cajoling creatures to take part in experiments. He has discovered that spider monkeys cannot resist Honey Nut Cheerios, that Asian elephants will do anything for a juicy carrot, and that if you're going to tempt a squirrel monkey, be sure to build an artificial tree bearing artificial nuts or it quickly loses interest. "If you leave them in a cage, they will loosen every screw in sight; they love to use their hands. So I devised fake nuts for this tree that they have to crack open. Half labeled with one scent contained a treat; half with another were empty." All seemed monkey-proof. "But to my dismay, they loved opening the nuts so much, they opened every single one! So I had to put them under time pressure. With a minute, they had to choose. They're quick, but thankfully not that quick."

Through experiments such as these, Laska has quantified the

smallest amounts of scent detectable to an animal. Time after time, on comparing the results to human studies, he has been surprised. "I have found that human subjects have lower olfactory detection thresholds. That is a higher sensitivity to the majority of odorants tested so far, compared to most mammal species tested so far." Humans, it seems, can do better than monkeys, macaques, sea otters, fruit bats, vampire bats, and even those we consider supersmellers, like mice, rats, and shrews. According to him, "Human olfactory inferiority is a myth."

The dog is the single species that Laska has found to flout this pattern. "Only fifteen scents have been tested on both humans and dogs, so only fifteen data points exist," he told me, "but it is interesting to note that humans even outperform the dog with five of the fifteen odorants. The five where we perform better are from plants, whereas seven of the ten where the dog retains an advantage are typical of its prey." Moreover, one of the five plant aromas is the ripe-banana scent of amyl acetate, the same chemical used in the oft-quoted pair of studies by James Walker to back up the dog's claim as the animal kingdom's superior nose. Laska discovered the existence of another study on human detection thresholds for amyl acetate; whereas Walker's found dogs can still smell it at concentrations as diffuse as −5.94 log parts per million, this study reports that humans have done so at, more diffuse still, −7.02 log parts per million. On hearing about Laska's work, Avery Gilbert asked mischievously, "Does that sound like across-the-board, doggy nose superiority to you?" To his mind, this proves that "the broader and much-cited claim that dogs are 10,000 to 100,000 times more sensitive to smells in general is unsubstantiated." In 2014, a study dislodged the hound's slipping crown further still.

The fragrance of the damask rose is an intoxicating mixture of over a hundred volatile components, but how many do we actually smell? When Linda Buck and Richard Axel won their Nobel Prize for olfaction, the press release claimed that their discovery "is the basis for our ability to recognize and form memories of approximately 10,000 different odors." Again Gilbert is skeptical of the total but respectful of its provenance, saying, "Surely that's a number we can take to the bank." It is, after all, also a number often heard in university lecture halls. That said, not everyone was convinced. "The figure came from a theoretical study, conducted in 1927, with questionable assumptions," the neurobiologist Andreas Keller told me. He cofounded the Rockefeller University Smell Study and put the theory to the test with living, breathing, and sensing subjects. He and colleagues waved hundreds of combinations of 128 different odors beneath the noses of untrained volunteers. They mixed ten, twenty, or thirty odors in jars, then deliberately made the differences between them smaller and smaller, hoping to find the point at which people failed to distinguish them. Then, using complex statistical techniques, they worked out our powers of scent discrimination. The final figure far exceeded expectations.

"Our results indicate several orders of magnitude more than thought," said Keller. "We estimate that humans can discriminate at least a trillion olfactory stimuli." In fact, the final calculation was 1.72 trillion. "This may seem like an astonishingly large number but because we only investigated mixtures of up to thirty components, this is probably the lower limit of our potential." Matthias Laska said, "I was so thrilled that someone had finally done the maths. If you read the old scientific paper that first came up with the 10,000 figure, it actually says that humans can smell 'at least' that number of different odorants. When the reference was cited, over and over, those two crucial words were missed." The Rockefeller study not only revealed that our sense of smell has been grossly undervalued but also upturned all notions of our sensory pecking order. Most auditory scientists would agree that an average human ear can discriminate between hundreds of thousands of audible tones. Most visual scientists would assert an unremarkable human eye can see several million shades of color. Yet our nose can smell at least a trillion different scents.

*

In 2017, a paper titled "Poor Human Olfaction Is a 19th-Century Myth" was published in the scientific journal *Science*. Its author, John McGann, a psychologist from Rutgers University in New Jersey, called into question the man who first described our sense of smell as impoverished. "Paul Broca classified humans as 'nonsmellers,' not owing to any sensory testing, but because of his beliefs," he told me. McGann realized that we are destined to fail if we take Broca's approach and compare what percentage of the brain is taken up by the olfactory bulb. "In the case of a mouse, its olfactory bulb occupies 2 percent of its brain, but ours is 0.01 percent of our brain," so mouse trumps man. "If you look at the absolute size, we actually rank rather well. Our olfactory bulb is fairly large, 60 cubed millimeters as opposed to 10 at most in the mouse." McGann pointed out that as our brain enlarged during evolution, our olfactory bulb might not have grown, but it did not shrink to make way for the frontal lobes or any other cerebral structure. However, his knockout blow to Broca comes from two recent reports by independent laboratories. They used a technique called isotopic fractionation to count the individual neurons in mammalian olfactory bulbs. "Both these studies show that size is in any case irrelevant. Mammalian olfactory bulbs, big and small, have the same number of neurons." Whether one is human, mouse, or star-nosed mole, broadly speaking, olfactory bulbs hold the same ten-million-odd neurons. So, given that the hardware of our nose is inferior to that of the dog—smaller olfactory epithelial patch, bearing fewer receptor types, combined with a less efficient sniff apparatus—what lends us the edge?

The Yale neuroscientist Gordon Shepherd recently suggested that we have fixated too simply on the olfactory bulb. "What matters more are the central olfactory brain regions that process the olfactory input," he explained, and these "are more extensive in humans than is usually realized." His list is certainly long: the olfactory cortex, olfactory tubercle, entorhinal cortex, parts of the amygdala, parts of the hypothalamus, the mediodorsal thalamus, the medial and lateral orbitofrontal cortex, and parts of the insula. "These regions enable humans to bring far more cognitive power to bear on odor discrimination," meaning "humans smell with bigger and better brains." Shepherd drew an analogy to the hair cells of the inner ear and language. He noted the scant differences in these sensory cell numbers between humans and rats, cats, or, as

Chapter 3 established, owls. "This modest increase in the input from the peripheral auditory receptors provides little basis for the development of human speech and language." Our linguistic skills owe more to our brain. Similarly our scent sensors cannot account for our expansive dictionary of smells. Perhaps our multifaceted brain more than makes up for our underachieving nose.

The latest scientific advances may salute our nose's surprising sensitivity and range, but the truth is that unlike Nick Carter and Bailey, we rarely value or rely on our sense of smell. In 2007, an eccentric experiment took place on the lush green lawns of the University of California, Berkeley, to reveal what happens when we do. Thirty-two students—half of them women, half of them men—signed up, only to get down on their hands and knees like a bloodhound. "Most people were game to try it," the neuroscientist Jess Porter recalled. "They just assumed, well, these are crazy scientists and this is another crazy thing that they do." Each human-hound was dressed in a blindfold, earmuffs, thick gloves, and kneepads to deaden sight, hearing, and touch—nearly all the senses except smell. Then they were asked to sniff out a 10-meter (30-foot) twisting path marked by twine that had been dipped in the essential oil of chocolate. The researchers proved that when we set our mind to smell, we are in fact quite capable. Not only are we rather adept scent trackers (two-thirds finished the course in under ten minutes) who improve with practice (those who repeated the exercise a fortnight later more than halved their times), but we naturally emulate dogs (zigzagging back and forth across the scent trail). Porter also found that success lies in having two nostrils. When she compromised some of her still-blindfolded volunteers by sealing off one nostril with tape, only a third finished the challenge and all took significantly longer. The results suggested that when we sniff, each nostril samples distinct regions and we compare their different incoming odor information. "This is analogous, we think," said Porter, "to the fact that you can use your paired ears to localize sound in space, and you can use paired eyes to have depth perception." The great gray owl's two ears give it earsight; our two nostrils give us stereoscopic smell. Although humans "don't use their noses to find a bakery in the morning," said Porter, "we've still retained an amazing ability to smell that can be revived with practice. That's exciting!"

✳

Three years before his death, Oliver Sacks made a reluctant confession. Stephen D., whom he had portrayed as a "highly successful young internist, a friend and colleague of mine in New York," was a complete fabrication. In fact, Sacks was describing himself; he had been the young man who had spent nights taking speed, cocaine, and PCP, only to awake one morning with the firm belief that he had become a dog. Sacks admitted to a journalist, "I'm less shy now. I think partly I'm at a distance from these things. They were forty years ago. And I don't think it's sensationalism or exhibitionism for its own sake—so much as the fact that I am basically constituted the same as everybody else, and I will get an inside take, as well as a scientific one." Maybe Sacks's canine metamorphosis had been drug induced, but knowing now how misjudged our noses have been, one can't help but wonder whether he was not so much in the throes of an olfactory hallucination as, perhaps more simply, an olfactory fascination. The late, great neurologist wrote that when imagining himself a dog, he had experienced a certain impulse to sniff and touch everything. Maybe Sacks's resulting sensory experience—"that smell-world, that world of redolence . . . so vivid, so real"—was neither imagined nor extraordinary. Perhaps that world is available to anyone who simply puts nose and mind to it. The tales told by Feynman, Laska, Keller, and Keller show that the dog beneath the skin exists within every one of us, waiting for our whistle.

8

The Giant Peacock of the Night
and Our Sense of Desire

ONE MAY EVENING IN 1877, IN A REMOTE CORNER OF Provence, a female moth known locally as the *grand paon de nuit*—giant peacock of the night, or *Saturnia pyri*—emerged slowly from her cocoon. Feathered antennae appeared, followed by a cream-capped head and filamentous legs on a downy body, all observed quietly by a man whom Victor Hugo would describe as the "Insects' Homer." Later, he would also be hailed as the father of entomology, but at the time, Jean-Henri Fabre was living in obscurity and near penury, seeking only the company of "his dear insects." The creature unfurled and spread her trembling taupe wings. Fabre wrote in *The Life of the Caterpillar*, "I forthwith cloister her, still damp with the humours of the hatching, under a wire-gauze bell-jar." He had little expectation of what might ensue, but with the rising of the moon, he was richly rewarded as visitors began flitting through the open windows of his house:

> With a soft flick-flack the great Moths fly around the bell-jar, alight, set off again, come back, fly up to the ceiling and down. They rush at the candle, putting it out with a stroke of their wings; they descend on our shoulders, clinging to our clothes, grazing our faces. The scene suggests a wizard's cave, with its whirl of Bats.

Fabre kept vigil, enthralled by the unfolding scene. "Coming from every direction and apprised I know not how, here are forty lovers eager to pay their respects to the marriageable bride born that morning." Over the next week he counted as many as 150 male suitors to his solitary captive; they even flocked to the bell jar after she had been removed. Fabre began to suspect that the males had been "apprised" by something on the air, something undetectable to his own nose. He had no way of knowing that these moths had been bewitched not by any ordinary scent but by an airborne aphrodisiac that would call into question what we mean by our sense of smell.

The giant peacock of the night is also known as the giant emperor moth or the great peacock moth. It lives up to the first parts of its name. With a wingspan approaching 15 centimeters (6 inches), it is the largest European moth. Its wild silk moth, or Saturniid, relatives are among the largest insects in the world, the behemoths of lepidopteran society. Although not colorful, the species caught the eye of Vincent van Gogh. He immortalized it in paint and, in a letter to his brother, described it as "astonishingly distinguished," noting "black, gray, white, shaded, and with glints of carmine or vaguely tending towards olive green." Males of the species exist to find and impregnate females. Without working mouths or digestive tracts, they cannot even take time out to eat. The clock starts ticking the moment they emerge from their pupae, as the food stores that fuel their missions will last but days. The female is rarely nearby, but her particular perfume ensnares males from afar; indeed, reports exist of one smitten Saturniid being lured from a distance of some 5 kilometers (3 miles). Three-quarters of a century after Fabre's "Great Peacock Evening," a Nobel Prize–winning organic chemist would also home in on this mysterious moth scent.

✳

Adolf Butenandt turned his attentions to a species more accessible than the giant peacock of the night: a domesticated cousin, known simply as the silk moth, whose cocoons are prized for their silken thread. In 1953, he bought all the specimens he could from commercial breeders and had them delivered to the Max Planck Institute in Tübingen. Carefully sorting through the pupae, preserving the males in methanol, he left the females to continue their hidden metamorphosis. Soon, Butenandt's laboratory was aflutter with newly hatched female silk moths. He set about slicing the tiny scent-laden glands, their *sacculi laterales*, from the tips of each abdomen and extracting the contents to distill their seductive ingredient. The work was delicate and painstaking, but he pioneered processes that can isolate minute quantities of molecules.

By 1956, he had nearly perfected his techniques and needed more subjects for his final experiments. Because the European silk industry had collapsed after the recent world war, he had to look farther afield and ended up importing half a million cocoons from Japan. The harvest from this vast number of eastern virgins was meager: a half-million sex glands produced no more than 12 milligrams, the weight of a small snowflake. Nevertheless, it was enough for him to divine the scent's molecular structure and synthesize it in the laboratory. By demonstrating that this molecular duplicate rendered males lovestruck, Butenandt proved he had identified the female moth's love potion. In 1959, he published his findings. He christened the (10E,12Z)-hexadeca-10,12-dien-1-ol molecule with the more pronounceable *bombykol* after the silk moth's Latin name, *Bombyx mori*. The same year, two colleagues, Peter Karlson and Martin Lüscher, coined a word for an odor like bombykol, which triggers certain behaviors in another of the same species. From the Greek words *pherein* (to transport) and *hormōn* (setting in motion), "a carrier of excitement," they came up with *pheromone*. Science is still revealing the full extent of this strange olfactory intoxication, laying bare its implications for our sense of smell.

As a sensory physiologist working today at the Max Planck Institute, Karl-Ernst Kaissling walks in Butenandt's footsteps. Now nearing retirement, he has spent the past five decades studying thousands of moths. Like his predecessor, he has them shipped from all four corners of the globe: "We get Bombyx from commercial breeders in Germany, England, France, Italy; Saturniids from Germany, the United States of America, and the Czech Republic. We very seldom grow some ourselves." Kaissling has released males into wind tunnels and watched what happens when traces of female pheromone are introduced. His work is precise, his mind particular; he told me, "We found that 50 percent of male silk moths responded with wing fluttering to a source load of about 2×10^{-5} micrograms with an air stream velocity of 60 centimeters [24 inches] per second." Suffice to say, male silk moths keep sensing pheromone until it is one or two parts per quadrillion. With antennae twice as large, the giant peacock moth is likely more sensitive still. The nose of the bloodhound may be legendary, but it deals in parts per trillion. The antennae of moths detect scents a thousand times more dilute; they sieve the passing air currents for sparse molecules of pheromone, leaving the dog sniffing aimlessly in their wake. However, the cross-species comparison creates a conundrum. "The numbers of olfactory neurons differ tremendously," explained Kaissling. "There are 34,000 bombykol receptors across Bombyx's antennae, but at least 100 million olfactory neurons in a dog's snout." The seemingly supernatural sensitivity of wild silk moth males is achieved with far fewer sensory receptors, an indication perhaps of the incredible potency of the female pheromone.

Using a device invented in the Max Planck laboratories known as an "electroantennogram," Kaissling has inserted wires into the male's thread-thin antennae to read nerves firing when pheromone-spiked air is blown across them. He has also isolated and studied the individual olfactory neurons. His experiments show that a single bombykol molecule alighting on an antennal hair's olfactory receptor is enough to trigger an impulse to the brain. "A much larger number of nerve impulses is needed in order to alert the moth," he explained. Even so, few molecules are sufficient to prompt the male to change his behavior, impelling him upwind in a turmoil of ardor toward the waiting female. The female pheromone is less chemical messenger than chemical command, and males yield to the faintest, most fleeting of encounters. Such knowledge unnerves male moths, but also us. In his essay "A Fear of Pheromones," the medical researcher and author Lewis Thomas wrote of the moth,

> It is doubtful if he has an awareness of being caught in an aerosol of chemical attractant. On the contrary, he probably finds suddenly that it has become an excellent day, the weather remarkably bracing, the time appropriate for a bit of exercise of the old wings, a brisk turn upwind. . . . Then, when he reaches his destination, it may seem to him the most extraordinary of coincidences, the greatest piece of luck: "Bless my soul, what have we here!"

Pheromones hint at a dark side to the sense of smell: a world of odorless odors and scentless scents that, once inhaled, work below radar to hijack another's behavior and strip them of free will. "Pheromones are not cognitively processed," stated the neuroscientist Rachel Herz, "but rather elicit instinctive and presumably unconscious responses." As the biologist Tristram Wyatt put it, "*Pheromone* is a very powerful word. It conjures up sex, abandon, loss of control." The possibility of human bombykol lands with a thud, threatening to explode our understanding of human relationships: of what really happens when we fall in love. Could it have us succumb like Titania to Puck's purple flower to dote madly on any passerby? Or, like the protagonist in Patrick Süskind's novel *Perfume*, to control others with an irresistible, if gruesome, master scent? Fiction aside, the hunt for human pheromones has been underway in earnest in university and commercial laboratories around the world. Military agencies have funded research into human pheromones, while the perfume industry has committed vast sums in a race to find and bottle the elixir of desire. The journal *Science* deemed the existence of human pheromones one of the outstanding questions of our time. Debate rages, often descending into such rancor that scientists now steer clear of the term *pheromone* when talking about our species unless it is prefaced with the word *putative*. Yet the evidence is tantalizing.

*

Research has shown that smell, more than any other sensory experience, has the ability to create and corral emotion. Rachel Herz described it as our most "emotionally connected sense." This is because smell has an intimate relationship not simply with the brain—a mere two synapses from the world outside—but with a particular part. Imaging studies indicate that whereas sight, sound, taste, and touch are first registered in the thalamus, smell bypasses this gateway and heads straight to the brain's emotional center, the amygdala. Herz said, "No other sensory system has this kind of privileged and direct access to the part of our brain that controls our emotions," adding, "I have often wondered whether we would have emotions if we did not have a sense of smell," and, with a nod to a certain French philosopher, "I smell, therefore I feel."

Fragrances conjure up the full sweep of emotions, from joy to rage; they can make us wistful, bring tears to our eyes, and even fill us with fear. "Have you ever been stricken with a feeling of dread, not known why, and then noticed a strange smell in the air?" asked Herz. "Thousands of New Yorkers had this exact experience walking in the streets and riding through the subway stops near the World Trade Center during the months after September 11, 2001. The strange charred and dusty scent was an instant reminder of that historical terror." Smells can also make our hearts ache; they can titillate or tempt. Quoting Vladimir Nabokov—"smells are surer than sights or sounds to make your heartstrings crack"—Herz has even rebranded olfaction "the amazing sense of desire." Such potency was not lost on our forebears.

Since the time of the ancient Egyptians, we have scoured the natural world and prized its perfumes for their erotic power. The well-known Victorian sexologist Havelock Ellis wrote that musk "has always had the reputation, more especially in the Mohammedan East, of being a sexual stimulant to men; 'the noblest of perfumes,' it is called in the *El Ktab*, 'and that which most provokes to venery.'" Musk—Sanskrit for "testicle"—is harvested from the glands of the Himalayan musk deer. Ellis observed, "Until recent times odors preferred by women have not been the most delicate or exquisite, but the strongest, the most animal, the most sexual." We have dabbed our necks and wrists with the oily yellow secretion from the anal glands of the civet cat, the waxy pathological growth from the intestines of sperm whales, and extracts from the anal glands of the beaver. These ingredients are still added to perfumes today. Our use of bestial odors exploits our brain's tight coupling of smell and emotion to quicken pulses, but our natural smell is reputedly more seductive still. In the sixteenth century, one nearly upset the French court.

In 1572, the Princess of Condé is said to have caught the eye, or rather the nose, of the future king of France. Yet the two did not even meet. It happened at a ball in the Louvre Palace during her wedding celebrations to another man. The nineteenth-century Parisian physician Charles Féré wrote, "Having danced a long time, and finding herself somewhat incommoded by the heat of the ball, this princess passed into a cloakroom." She changed and left her damp clothes behind, but when the duc d'Anjou, the man who would become Henri III, entered to brush his hair, he wiped his face by mistake with the princess's chemise. From that moment, as Féré noted, he "conceived for her the most violent passion."

Henri is but one in a long line of men throughout history who have reputedly swooned before the scent of a woman. Recall Napoleon's notorious words to Empress Josephine—"Home in three days. Don't wash!"—or those of the nineteenth-century French novelist Joris-Karl Huysmans on how the aroma of a woman's underarms "easily uncages the animal in man." The *Kama Sutra*, written by men, claims that a woman's beauty lies not in her looks but in how she smells. Elizabethan ladies allegedly peeled apples to put beneath their armpits, removing them when saturated with sweat and gifting these "love apples" to their sweethearts. A similar tradition filtered down the generations in rural Austria, where, until recently, girls would keep a slice of apple under their arm at dances, before presenting it to their chosen partner to eat. History is brimful of anecdotes attesting to the allure of a woman's unadorned scent, but now so too is science. Possible evidence for this has emerged from an unlikely subject of academic study: the gentlemen's clubs of Albuquerque.

Geoffrey Miller, an evolutionary psychologist at the University of New Mexico, became curious about the scientific potential of lap-dancing bars. He wondered if, by enabling men to freely choose women, they might provide a window into the male psyche and the rules of attraction. In an issue of *Evolution and Human Behavior* published in 2007, Miller wrote, "Lap dances are the most intimate form of sex work that is legal in most American cities. Patrons can assess the relative attractiveness of different women through intimate verbal, visual, tactile, and olfactory interaction, and those attractiveness judgments can directly influence women's tip earnings." This monetary transaction is, moreover, a clear measure of male desire and, as Miller claimed, the first real-world economic evidence of male sensitivity to female attractiveness. The team recruited eighteen exotic dancers, then asked them to record their earnings over two months and some 5,300 dances. Miller pointed out that the researchers never even needed to set foot in the strip club; the data were gathered via a website.

The scientists found that female allure shifted through the month, waxing and waning with the women's reproductive cycles. The dancers were earning almost twice as much in tips—taking home on average $334 as opposed to $184 per night—when they were ovulating and fertile. Miller became curious to understand how the women signal, and the men sense, the change. He observed, "Tip earnings are unlikely to be influenced by cycle shifts in stage-dance moves, clothing, or initial conversational content because these cues just do not vary much." He proposed a cue that flies "below the radar of conscious intention or perception": one that the women are unaware of sending and the men oblivious to receiving. It could be airborne. The study was chosen by both the organic chemist George Preti and Rachel Herz as a potential example of a scentless pheromone casting its spell. Without further evidence, as Herz admitted, "The cause of this provocative finding is still a mystery." Whether men are as susceptible as male giant peacock moths to female pheromonal subterfuge remains unresolved. On the question of where such a putative pheromone might be made, scientists are on surer ground.

*

In *The Scented Ape*, the zoologist David Michael Stoddart claimed, "Man has to be considered as quite by far the most highly scented ape of all." As we do not brandish moth-like scent glands on our tail end, Stoddart's focus falls on our skin and the unusually high numbers of glands that it conceals. Sebaceous glands exude a thick, oily substance thought to keep the skin moist and supple. Two other types make sweat. Eccrine glands, found all over our body, are the volume producers; some 3 million of them release up to 3 liters (5.25 pints) of watery solution during every hour of exercise to cool us down and regulate body temperature. By contrast, apocrine glands—restricted to Huysmans's "spice boxes," our armpits, and other hairy parts—are stimulated to sweat only in stressful, or exciting, situations. They become functional at puberty and do so differently between the sexes, being larger and more numerous in men. When their complex chemical secretions are broken down by our skin's bacterial microflora into smaller and lighter molecules, they waft into the air and give rise to our body's natural odor. Consequently, it is generally thought these glands could release pheromones. As Tristram Wyatt said, "We give off hundreds, if not thousands, of different volatile molecules, so pheromones might be hidden among this cloud of odor molecules." Analyses of male and female sweat have found more androgen-related steroids in the former. This includes one long recognized as the sex pheromone of pigs. Unlike the moth's love drug, androstenone is released in the saliva of the male pigs as a "come hither" to the females. It is so effective that it is canned commercially as Boar-Mate, with the guarantee that a few sprays make any sow receptive to any boar's advance. This discovery turned the tables on the mating game. A slew of scientific studies were undertaken to ascertain whether these androgen steroids—androstenone, androstenol, androstadienone, and their ilk—could be mankind's answer to bombykol.

The most famous experiment took place in a dentist's waiting room. Scientists engaged the help of a local clinic. Every morning for three weeks, they sprayed the back of one of the twelve seats with androstenone and waited for the patients to arrive. They found that women were drawn significantly more to the spiked seat; the pattern recurred despite diminishing concentrations to as little as 3.2 micrograms. The researchers wondered whether the women still perceived the odor. A similar experiment was repeated more recently, using homosexual men as well as heterosexual women, but this time the scientists conducted threshold tests to ensure few, if any, would notice a smell and only when standing next to the chair. They discovered that heterosexual women, again, and homosexual men gravitated toward the sprayed seat. Another study stealthily administered androstenol to women through necklaces. "Student volunteers were exposed unknowingly overnight to the vapor of [the] pheromonally active substance and compared with controls." After wearing either pheromone or pheromone-free necklaces overnight, they recorded their conversations the following day. The scientists learned that those wearing androstenol recalled more, longer, and deeper interactions with men, and proposed that the pheromone had put these women in the mood for love.

Over the past few decades, androgen steroids have been put to the test in all manner of imaginative ways. Under their influence, subjects have been asked to rate photographs for attraction and CVs for job suitability, their moods have been gauged, and their brains scanned. Then in 2008, for the first time, a study placed people in a real-life romantic scenario: straight into the cut-and-thrust of the singles circuit and the latest trend for rapid-fire courtship.

Speed dating was the brainchild of an Orthodox rabbi looking for ways singles could meet and marry, but what started at Peet's Coffee & Tea in Beverly Hills soon went global. Typically, women remain seated at a table while men circulate. Hopefuls are given three minutes to sell themselves before the bell rings to move on. At the end, preferred choices are fed back to the organizers; if there is a match, numbers are swapped. "Context is key so I had reservations about the lab-based attraction studies," psychologist Tamsin Saxton told me. "We wanted to do an ecologically valid study to see if these androgen steroids make a difference in the real world. So, speed dating was perfect." Saxton, at the University of Liverpool at the time, organized the event at the student union on campus. To ensure the students did not know one another, she sourced the men from two nearby universities. She recalled, "I spent days running around, inviting students. I was nervous I wouldn't get enough people; speed dating can be intimidating, but the turnout was great and people made a real effort dressing up. Perhaps it was the offer of the free drink at the end of the event that swung it!"

Students were under strict instructions not to drink before the event and not to wear any strongly scented perfumes, aftershaves, or deodorants. The female students were split into three groups: the skin beneath their noses was wiped with water, clove oil, or androstadienone diluted in clove oil. Saxton explained, "We used concentrations of androstadienone thought to be below conscious detection levels for the majority of the population, but added clove oil to mask any scent just in case." With the women seated, the first bell sounded, and the dates started. "I don't know whether any marriages were made that evening," said Saxton with a smile, "but we got interesting results." Men were rated as more attractive when assessed by women who had been exposed to androstadienone. "We decided to repeat the exercise with older participants. Thankfully, these were organized for us through a private speed-dating agency at a local bar." However, only one of the two further studies supported the first finding. Saxton is the first to admit that more studies are needed, "But we found that the androstadienone effects on two groups of women were strong enough that the combined results were still significant. Perhaps these putative pheromones don't dictate so much as subtly modulate behavior." Her research shows that a steroid in a man's armpits might act subconsciously on a woman's mind—not controlling, perhaps, so much as subliminally swaying whom, if anyone, she takes home.

The findings of these and many other androgen steroid studies are suggestive, but none is beyond criticism: each is accused of either flaws in experimental design or an inability to replicate the results, sometimes by the experimenters themselves. Moreover, there are just as many studies where androgen steroids failed to act as pheromones: failed either to work subconsciously or to elicit the predicted behavior. In his book *The Great Pheromone Myth*, olfaction scientist Richard Doty ridiculed the entire scientific enterprise by asking, "Are women, in fact, attracted to the odors of male pigs or more willing to have sex in the presence of such odors? Are birth rates or other indices of sexual behavior higher in states or counties with pig farms?" He goes further. Citing Lewis Carroll's "The Hunting of the Snark," a poem that recounts "the impossible voyage of an improbable crew to find an inconceivable creature," he likened the quest to find human pheromones to a snark hunt, adding that "pheromonology has become the modern phrenology of chemosensory science." Nevertheless, other scientists refute this, proposing that there is evidence that we demonstrate at least one pheromonal response, albeit one that does not directly involve men.

*

"I was the only woman, and the only undergraduate, in the room. I remember feeling deeply embarrassed and blushing as I raised my hand," recalled Martha McClintock. Now the founding director of the Institute for Mind and Body at the University of Chicago, she was a student at Wellesley College in summer 1968 and attending a seminar packed with many eminent biologists. They were discussing mice and how the females use pheromones. Whereas female giant peacock moths use pheromones to effect a change in the male's behavior, female mice use them to change another female's physiology and synchronize their reproductive cycles. McClintock lived in a dormitory of over one hundred women and, as the person designated to fetch feminine hygiene products, she had noticed a peculiar pattern. "Once a month people would start yelling 'We're out!' and I would get on my bike and ride down to the drugstore." At the seminar, she stood up and spoke out in front of a mainly male audience. "I made the point that women living together also had synchronized menstrual cycles. I can then remember all of these eyes staring at me," she recalled. "I got the impression that they thought it was ridiculous. But they had the courtesy to frame their skepticism as a scientific question: 'What is your proof?'" And so began one of the most famous experiments in human olfaction.

Enlisting the help of dorm mates, McClintock set about collecting evidence for her observation across the following academic year. One hundred thirty-five females were asked when their cycles started, how long they lasted, their room number, and with whom they spent the most time. McClintock discovered that at the start of a new term, everyone's cycles differed, but after four months, they shifted to follow those with whom they had most contact. Her professor, the great sociobiologist Edward O. Wilson, encouraged her to publish; when her paper was accepted by the illustrious British journal *Nature*, she was only twenty-two. In it, she stated, "Although this is a preliminary study, the evidence for synchrony and suppression of the menstrual cycle is quite strong. . . . Perhaps at least one female pheromone affects the timing of other female menstrual cycles."

It took nearly three more decades to unravel the biochemistry behind her observations. In 1998, again in *Nature*, she reported the discovery of two odorless compounds from female underarm sweat that work in tandem on other women. The research revealed that when sweat produced before ovulation is applied to another woman's upper lip, it acts to accelerate the recipient's ovulation and shorten her cycle, whereas sweat produced at ovulation acts to delay another's ovulation and lengthen her cycle. This chemical coupling enables a woman to effect a physiological change in another, exerting control over their fertility without their awareness. Consequently and to great fanfare, McClintock was able to claim that her study at last "provides definitive evidence of human pheromones." Yet like the androgen studies, the results have been met with skepticism, with many scientists, not only Richard Doty, still in need of persuasion.

*

The Oxford University zoologist Tristram Wyatt said wryly that the controversy over human pheromones remains "almost as heated as the 'stink wars' between opposing troops of ring-tailed lemurs, which wave their pheromone-coated tails to assert their dominance." He told me, "If only we knew as much about humans as we do about moths." Research since Adolf Butenandt has shown that pheromones come in various guises. Those that control behavior, like the male summons wafted onto the night air by a female giant peacock moth, are now called re-leaser pheromones. Those that advertise information such as fertility and sway behavior, seen possibly in Geoffrey Miller's exotic dancers, are signaler pheromones. Those that subtly alter mood or libido, at work perhaps in Tamsin Saxton's students, are modulator pheromones.

Finally, those that produce physiological responses, found maybe in Martha McClintock's sweat, are primer pheromones. These definitions, just like the experimental conclusions, are not without contention. "We need to start again," Wyatt said, and he means this literally, with a return to Butenandt: "He created the model for how we should go about pheromone analysis." By isolating the female moth's active ingredient, synthesizing it in his laboratory, then proving that his man-made version attracted male moths, "Butenandt closed the circle and that's the thing that has never been done in humans." Wyatt believes that McClintock and other pheromone hunters must do the same in order to erase the word *putative* and prove beyond doubt the existence of human pheromones for love, fertility, and more. Even so, he is encouraged: his evidence is the natural world.

The field has seen a spate of pheromone discoveries since Butenandt's 1959 revelation. "Pheromones have been found across the animal kingdom, sending messages between courting lobsters, alarmed aphids, suckling rabbit pups, mound-building termites, and trail-following ants," observed Wyatt. "They are also used by algae, yeast, ciliates, and bacteria." The female giant peacock of the night is far from alone in releasing love lures—so too do over a hundred other female moths and butterflies—but in monarch butterflies, it is the males that coat females with a love dust of fine, pheromone-impregnated particles. Studies have also found that male fish—from gobies to Mediterranean peacock blennies—use pheromones to entice females. Certain species "eavesdrop" on these chemical communications. "When researchers in California put out traps to field-test bark beetle aggregation pheromones, they were surprised to also catch 600,000 predatory beetles. It was later found that the predators had evolved olfactory receptors highly sensitive to their prey's pheromones." Other species make counterfeits. The female American bolas spider spins large, sticky, ball-shaped lures that she covers with fake moth pheromones and swings at the duped incoming male moth. If the bolas strikes, the moth rarely escapes. Wyatt added, "She produces different pheromone blends through the night to match the different peak flight times of her victim moth species."

Mammals were the last animal group to have their pheromones unveiled. "Most spectacular was the 1996 discovery of the female Asian elephant's sex pheromone, a small molecule also used by some 140 species of moth as a component of their female sex pheromones." All in all, "pheromones have been identified in almost every animal you can think of." Wyatt asked, If the entire kingdom is awash with airborne aphrodisiacs, why should we be any different? "As we're mammals, we are likely to use pheromones," he concluded. "There may never be a magic potion to make us irresistible, but I'm sure human pheromones will surprise us yet." In fact, scientists are already on the verge of a breakthrough: one that promises to close Wyatt's circle of proof not in couples but in what comes from their coupling.

Research at a French hospital's maternity ward has shown that hours-old babies will unmistakably and reliably respond to a scent from their mother's nipple. They wake up, turn their head, open their mouth, stick out their tongue, and start suckling. Scientists argue that these first few hours of our lives are the most dangerous: if a baby fails to find milk, it will die. A pheromone released by the mother to induce her baby to find food seems the perfect solution. Indeed, the team, led by Benoist Schaal, has already isolated such a mammary pheromone in rabbits and now hopes to isolate it in humans. The chemical compound in the mother's secretions fulfills various key criteria. First, as Schaal said, "This finding is especially interesting because such a stimulus is virtually undetectable to adult humans." Second, it prompts an instant and obvious reaction. Wyatt has seen what happens when a glass rod dipped in the maternal secretion is waved beneath the nose of a sleeping babe. "It's a connoisseur's reaction of delight," he said. "It opens its mouth and sticks out its tongue and starts to suck." The baby will also respond to secretions from another mother: "It's not about individual recognition. It can be from any mother so it really could be a pheromone."

This potent but indiscernible airborne message could trigger suckling behavior across our species. Wyatt told me the aim is to follow Butenandt's protocol: "The French team are being rightfully cautious, but if we could identify this molecule and synthesize it, we could then use it to make premature babies more likely to suckle and survive." They would also at long last have incontestable evidence of a human love pheromone, albeit with a twist, between a mother and her child. While the hunt for the more conventional lust-inducing philter continues, evidence has emerged exposing other ways in which we subconsciously succumb to smell when looking for love.

*

Sleep in a T-shirt for three days, bag it, and bring it along to a bar. "Pheromone parties," which began in New York in 2010, have already crossed continents and oceans; London hosted its first in 2014. On arrival, partygoers leave their T-shirt in a numbered bag on a table. Throughout the night, it and the other T-shirts are sniffed. When smell preferences coincide, phone numbers are swapped. "I'm an evolution-ary biologist, not an odor scientist," said the man behind the craze. "I'm interested in odors only from the perspective of mate choice." Claus Wedekind is now a professor in evolution and ecology at the University of Lausanne, but two decades ago, as a student at Bern, he was casting around for a dissertation topic. A late-night stint of research offered in-spiration. "I was in the library. It was dark outside, and most people had left for the evening, when I found this article about how mice choose mates based on their MHC." Like Martha McClintock, Wedekind de-cided to see if what is true of mice is also true of humans. "I was so excited by the thought that I couldn't sleep that night." MHC stands for major histocompatibility complex: a cluster of genes that code for our immune system. Everyone other than identical twins possesses a set of MHC genes as individual as a fingerprint. To establish whether this is somehow manifested and then detected in humans through odor, We-dekind thought up a new experimental protocol. He told me, "T-shirt sniffing seemed like a good way to tease out scent from all the other signals sent and received during courtship." Little did he know that this would capture imaginations around the world.

Wedekind's priority was to find some willing sweat donors and even more willing sniffers. Like Saxton and McClintock, he looked to his university. "I targeted different departments so the subjects wouldn't know each other." He soon had fifty male sweat givers and fifty female receivers, but then disaster struck. "Journalists got hold of the story and then politicians. One saw the words 'human, mate choice, genetics' in my proposal, tried to shut down the project and get me expelled. Then she gave a nasty interview for an important local paper. Some of my volunteers read this and dropped out." Other students started a protest. Newspapers quoted one who insisted the study was unsuitable because it "optimized" offspring and another who claimed the study demeaned women by implying their "functions and abilities are reduced to reproduction." The furor grew. A politician denounced the project as "Nazi research," and Wedekind's superiors at the university were soon embroiled in the controversy. Thankfully, the university president supported the research, but by then, it had been set back.

Six months later, with a somewhat smaller group of volunteers assembled, the experiment got underway. The research team took blood samples from the whole group and sent them off for serological analysis. This gave Wedekind everyone's MHC gene profile. He instructed the women to look after their sense of smell by using a nasal spray for fourteen days and read Süskind's *Perfume* to make them more mindful of their olfactory perception. He told the men to refrain from anything that might alter their body scent: whether sex, smoking, alcohol, even entering a smelly room. During the delay, chemists had been busy designing special soaps and detergents that cleaned without fragrance to use when showering and washing bedclothes. Finally, he gave the men a T-shirt in a plastic zippered bag and asked them to sleep in it for two nights running. "On the morning of the third day, I collected the T-shirts," Wedekind recalled, "but I found that the odors were really very faint; I could hardly smell them. I was terrified. A lot was riding on these experiments. I had received all sorts of support, and it looked like it would be a huge embarrassment."

That afternoon, he presented the men's T-shirts in boxes to the women. With noses snuggly slotted into a triangle cutout, they rated what they inhaled for attractiveness. "Thankfully the boxes amplified the smell a little, but it was still very faint, so I told the women it would all be in the first sniff. If you miss it on the first sniff, it would be too late." The experiment was repeated over the course of a few weeks as freshly worn T-shirts arrived. "I wanted all the women to be in the second week of their menstrual cycle, just before ovulation, as research had shown that smell sensitivity peaks at this time." With the last woman tested, Wedekind started to analyze the data. "This was the first time I had done an experiment on humans. So many of my past experiments on animals had not worked that I could not believe the results at first," remembered Wedekind. "Everything I had tested was significant."

Despite the near-imperceptibility of the body odors, Wedekind found the women were using them to make decisions. There was a clear pattern in their choices when considered alongside the MHC profiling data. "I realized that no one smells good to everybody. It depends on who is sniffing whom, and it is related to their respective MHCs." Women tended to rate men more attractive when there was a difference between their MHC genes. Like female mice, they were unconsciously favoring partners whose genes complemented their own. Also they claimed their favorites often reminded them of boyfriends past or present. "These memory associations were the most interesting piece of evidence," he told me. "They gave me the courage to suggest that this odor preference is happening in the real world. They are why I dared to use the words 'mate preferences in humans' in the title for my paper." However, his excitement was short-lived; not only did he find it difficult to get the paper published but also, when it finally reached press, the response of the scientific community was far from positive. "At the time so many people could not accept the result. It was really frustrating. I kept asking—why are you ignoring this? I now think people just couldn't believe it." Support for Wedekind's theory would come from a small corner of America that is "forever Europe."

The Hutterites, a self-sufficient religious community in North America, were originally established in 1528 in the Tyrolean Alps. Persecuted, they fled, arriving and settling in South Dakota in the 1870s.

Since then they have thrived; the three original colonies now number around 350, and the 400 founders have become over 35,000. Hutterites have begotten Hutterites. Most marry individuals from different colonies; the men travel to help farm the fields at harvest, the women to help with housework or gardening or to attend to a sister who has just given birth. Marriages are not prescribed, and romance is allowed to take its course, but given everyone is descended from so few—what scientists call a "genetic bottleneck"—it is surprising that hereditary diseases associated with inbreeding are not more common. Carole Ober, a geneticist from the University of Chicago, studied the Hutterites for over a decade and discovered that marriages do not occur randomly but between those with the most different MHC. Her research uncovered an experiment that had been taking place in nature for over a century. Its results affirmed those of Wedekind. "I was so excited to learn of Ober's work. I read it carefully and was completely convinced," he told me. "It was about time our findings were being taken seriously."

Wedekind's original investigation has since spawned more research and much debate. "At the time, people could not accept the result. It's been a long fight, but it's now a textbook study," he said. His discovery hints at the real body chemistry at work, and at play, in courtship. The cloud of molecules that we each release is unique and replete with messages. "The odors are so faint that you might not notice them at first. I suspect they are not important when you meet someone, say in a bar, but over the next week or month, you start to sense whether you like them or not." They operate at the edge of perception, and we remain unaware of their influence over us. You may think you are drawn to the tall, dark, and handsome stranger or the buxom blonde with fluttering eyelashes, but attraction of the nose may overrule that of the eye, and its preference is for opposites. Wedekind's work shows that we are drawn to those whose smell signifies genes that complement our own: evolution has honed matchmaking to create healthy babies with robust immunities. Although these odors work on us subliminally, manipulating whom we find attractive, they are not pheromones because they vary between individuals. "Our sense of smell is so much more complicated than anyone could imagine," said Wedekind. "Süskind in his book *Perfume* was wrong to assume that there is a perfect body odor."

✳

The bloodhound exemplifies how we consciously smell the world around us. The moth exposes an unknown side to this known sense: how we subconsciously smell our partners in love. In *Letters to a Young Poet*, Rainer Maria Rilke wrote, "For one human being to love another is perhaps the most difficult task of all, the epitome, the ultimate test." Little did he know, we love and lust in ways that we neither understand nor control. Attraction is guided perhaps by pheromones but certainly by odors. "People are skeptical because it goes on 'right under our nose,'" Claus Wedekind told me, while Martha McClintock suggested such knowledge is intimidating because "when it comes to sex and love, people really want to believe they are in control." The potent promise of subconscious smell extends beyond the sphere of desire. "Pheromones are not just about sex," said Tristram Wyatt, listing how animal pheromones influence families and friendships, as well as induce fear and flight. "There could be all sorts of things that humans do with pheromones that we simply don't know at the moment."

Ultimately, the field of pheromones and other subliminal smells has sparked controversy because it poses fundamental questions about all aspects of our lives: how we think, how we feel, even whether we have free will. When Jean-Henri Fabre died, the *New York Times* mourned his loss with an article on October 11, 1915. His words guide us from beyond the grave: "Because I have stirred a few grains of sand on the shore, am I in a position to know the depths of the ocean? Life has unfathomable secrets. Human knowledge will be erased from the archives of the world before we possess the last word that the gnat has to say to us." The same goes for the last word that his giant peacock of the night has to say about our sense of smell and its secret dimensions.

9

The Cheetah and Our
Sense of Balance

A T THE BERLIN OLYMPICS IN 2009, USAIN BOLT CROSSED the finish line of the 100-meter sprint in a breathtaking 9.58 seconds to become the fastest human in history. Three years later, on a sunny summer's day, a cheetah called Sara from the Cincinnati Zoo made it into the Guinness World Records by shaving seconds off the Lightning Bolt's performance. She blazed the 100 meters on a USA Track & Field–certified course in 5.95 seconds. One onlooker described her as "a polka-dotted missile." "Nobody can run like Sara," added her keeper. "I always knew she could run under six seconds, but to see it happen like this is wonderful." Light, with a small head and slender body, the cheetah, *Acinonyx jubatus*, is built like an arrow and, at full throttle, essentially airborne. Its spine—the longest and most flexible of all the felines—acts like a coiled spring, sweeping legs forward with 7-meter (23-foot) strides. Half of its mass is muscle, mostly the fast-twitch type that prioritizes rapidity over endurance. Consequently, cheetahs like Sara explode out of the starting block, going from 0 to 95 kilometers (59 miles) per hour in under 3 seconds. Not only is Usain Bolt left in a cloud of dust; so too are greyhounds and racehorses, animals bred for speed. Only a supercar like the LaFerrari Aperta, recently unveiled to mark the company's seventieth anniversary, can match its extreme accelerations. The cheetah is nature's ultimate land speed machine. Yet in 2013, the first motion study of it hunting in the wild revealed that speed is not its strength.

Alan Wilson, of London's Royal Veterinary College, is one of the world's foremost experts on cheetah locomotion. He runs the Structure and Motion Laboratory, where he analyzes captive cheetahs with high-speed cameras and medical scanners. A recent venture took him to the plains of Kenya. "We had only one valid and ratified measurement of the cheetah's maximum speed," he told me. "This was a cracking paper by Craig Sharp that demonstrated speeds of 60 miles [just under 100 kilometers] per hour." Sharp, who held the best time for running up Tanzania's Mount Kilimanjaro, had been the physiologist to the British

Olympic team. However, the study had been done in 1965 with a semi-tame cheetah. As Wilson explained, "It was properly old school, involving a Land Rover, some string, a piece of meat, and a stop clock." Given it was over only a 200-meter (220-yard) course, Wilson wondered whether in the wild, cheetahs might simply continue accelerating over longer distances and even improve on Sharp's record.

Wilson had at his disposal some purpose-built and highly innovative tracking collars. Equipped with GPS and electronic motion sensors, they record all manner of information other than speed, including acceleration, deceleration, turning forces, and banking angles. On arriving in Kenya, the research team fitted them to five wild cheetahs. Over the next eighteen months, they left the two males and three females alone, safe in the knowledge that the collars were monitoring every moment, day and night, as they ranged, played, fed, and hunted. Back in England, as the researchers sifted through the vast amounts of data, they saw some surprising patterns. "When you set out to do science, you have few expectations. The only thing we knew was that the animal had shown remarkable speeds, but that was the one thing we didn't see," said Wilson. "Our top speed was only 93 kilometers [58 miles] per hour, and most hunts were in fact half this." The cheetahs were not clocking record speeds; the evidence pointed toward other skills.

"We had some of the highest measured values for any land animal for both forward acceleration and deceleration," said Wilson. In a single stride, the cheetahs were speeding up by 3 meters (9.8 feet) per second each second and slowing down more dramatically. This degree of speed control had not been seen before, not even in animals selectively bred to be agile. "We saw acceleration and deceleration values almost double those published for polo horses." Further data revealed that these accelerations also continued through the sharpest of banked turns. "The lateral accelerations were very high—13 meters [42 feet] per second per second—again more than twice that of a polo horse, again one of the fastest recordings we have for a land animal." Cheetahs hunt gazelles, springbok, gemsbok, waterbuck, bushbuck, and impala. These antelopes may be quick, but they are effortlessly outstripped by a cheetah in a straightforward running race. They are better known for their light-footedness and their talent for turning tail. "We have always thought of cheetahs as sprinters, but now it looks as though this is only part of the story." Wilson concluded, "Like any athlete, the cheetah can run fast, but in the wild, it tends not to. Instead we discovered that successful hunts require exceptional maneuverability." To close in on the heels of their jinking prey, cheetahs like Sara from Cincinnati have redefined life in the fast lane; not only are they the undisputed champions of land speed but also they flaunt remarkable agility. This winning, if lethal, combination is made possible by a sense so underappreciated that to most, it remains wholly secret.

✳

In *The Man Who Mistook His Wife for a Hat*, Oliver Sacks narrated meeting an elderly gentleman with a rather unusual problem. "Others keep telling me I lean to the side: 'You're the Leaning Tower of Pisa,' they say. 'A bit more tilt and you'll topple over.' " The slant had been obvious to Sacks from the moment his patient had walked into the clinic. He was so far off to the left—about twenty degrees—that the gentlest of nudges would surely topple him. Yet Mr. MacGregor was convinced that he was plumb vertical, adding, "I feel fine. I don't know what they mean. How could I be tilted without knowing I was?" Sacks decided to film him walking. On seeing the footage, his patient became profoundly shocked. "I'll be damned!" he exclaimed. "They're right, I am over to one side. I see it clear enough, but I've no sense of it. I don't feel it." Mr. MacGregor was suffering from a disturbance to a sense we each possess but whose absence is more apparent than its presence because none of us feels it.

"We have five senses in which we glory and which we recognize and celebrate, senses that constitute the sensible world for us. But there are other senses—secret senses, sixth senses, if you will—equally vital but unrecognized, and unlauded," Sacks told Mr. MacGregor. These work below our conscious radar. Sometimes their sensory organs are associated with known senses; sometimes they are hidden from plain sight.

Consequently, they eluded not only Aristotle but also many of the great minds that followed. The sense that had gone awry in Mr. MacGregor had not been uncovered until the early nineteenth century, when an enterprising Frenchman called Flourens looked at pigeons. It took a further half-century before its significance in humans was appreciated. As Sacks's patient struggled to understand his particular affliction, he asked, "There should be some feeling, a clear signal, but it's not there, right?" Mr. MacGregor turned to his former trade—carpentry—and added, "We would always use a spirit level to tell whether a surface was level or not, or whether it was tilted from the vertical or not. Is there a sort of spirit level of the brain?" He would have been hard-pressed to land on a more apt analogy.

Locked deep in our ears, within the fabric of our skull, are two bony labyrinths. These inner ears are no bigger than garden peas. Yet each contains the snail-like cochlea, where sound waves start their journey into hearing, as well as two further kinds of sensory organ. The semicircular canals and the otolith organs are what keep us standing upright, head held high. Together they form our vestibular apparatus and endow us with a sense of balance or, from the Latin *capere* (to grasp) and *equilibrium* (balance), equilibrioception. Within their bony casings are membranous sacs holding substances that act like the spirit in Mr. MacGregor's level, as well as highly sensitive hair cells that fire signals to the brain. The otolith hairs are embedded in a gelatinous substance within a membrane, whereas those of the canals are submerged in viscous fluid called endolymph. Crucially, when our head moves, both membrane and endolymph follow more slowly and this small inertial lag bends the hairs, intensifying their electrical chatter to the brain. As the neuroscientist Brian Day told me, "It is this 'getting-left-behindness' that the inner ear measures." If our ears are full of seawater, we feel the effect of gravity as we move our head around, but our balance spirit levels operate silently within us. The sensory hairs keep up a continual monologue that makes our vestibular sensation different from most other senses, producing no recognizable, localizable, or conscious sensation. Although this sense may seem mute, it works ceaselessly.

We have ten natural spirit-level organs; each labyrinthine inner ear has two otolith organs and three semicircular canals. The otoliths respond to changes of speed in a straight line, whether accelerations or decelerations. They are positioned so that one is sensitive to the vertical ups and downs of an elevator, the other to the horizontal hurtle of a car. By contrast, the semicircular canals register all the rotational speed changes in three different planes through all the undulations and spins of a fairground Tilt-A-Whirl. Positioned at ninety degrees to one another, the lateral canal runs horizontally, roughly parallel to our chin, whereas the other two are vertically arranged so the anterior canal is angled forward, its posterior partner backward. Nod your head in agreement, in a movement technically known as pitch; both anterior canals of the right and left inner ear will fire up, whereas their posterior partners will quiet down. Bring your ear to your right shoulder; such a roll will excite both the anterior and posterior canals in the right ear and calm those same canals in the left. Finally, turn your head side to side in disagreement and this yaw is mainly detected by the lateral canals. Ultimately the brain integrates the varying degrees of pitch, roll, and yaw from the firing patterns of all three canals and two otoliths, across both ears; only then can it approximate where the body exists in space. So if we trip, the acceleration of our head falling activates the inner ear balance organs, alerting our brain to take action to regain stability. Scientists now know that rudimentary spirit levels evolved in deep time, so even the earliest life on earth could work out which way was up. Labyrinthine canals and otolith organs arose with the origin of the vertebrates some half a million years ago. Today, variations on the theme exist in all backboned animals, some more extreme than others.

*

"I've always been fascinated by the cheetah—the beauty of its movement, its speed and agility. Many studies had been done on its locomotion but no one had ever investigated its inner ear and how it achieves such balance," said Camille Grohé. In 2018, she was doing paleontological research at the perfect place to ask such a question.

The American Museum of Natural History houses 32 million specimens and artifacts, rising by some 90,000 each year, so boasts one of
the world's most encyclopedic research collections, including the odd
cheetah skull. "When you work in the museum, you are allowed to
borrow specimens like books from a lending library," Grohé told me.
"I recall my first trip into the mammal collection. Corridors, rooms,
and rooms, all lined with fossil cabinets from floor to ceiling. I was so
overwhelmed, I walked and walked and lost track of time." Eventually
she found the cheetah drawers and selected seven intact skulls. To
understand whether they were specialized, she also took specimens
from every major lineage of the cat family: a tiger, a leopard, a clouded
leopard, a fishing cat, a marbled cat, an African golden cat, a margay,
a bobcat, an African wildcat, a cougar, a jaguarundi, and—last but not
least—a domesticated house cat. With this unusual cache assembled,
she wheeled the crania underground into the bowels of the museum.
"The scanning facility is down in the basement, and I was headed for
the CT scanner." Scientists had used X-ray computerized tomography
to see inside dinosaur eggs, meteorites fallen from space, even Egyp-
tian mummies; Grohé was the first to use it to look through bone into
the inner ear of a cheetah.

With the skulls recast as three-dimensional virtual models on
Grohé's computer, the precise size and morphology of their bony labyrinths were laid bare. At first glance, the cheetah's seemed like another
variation within the feline lineup. She digitally mapped each inner ear
and teased apart the spaces dedicated to balance from those used for
hearing: the semicircular canals and otolith organs from the spiraled
cochlea. Next, she plotted these vestibular volumes against body mass.
"Throughout nature, we find larger balance organs in larger animals, so
this enabled me to search for other changes," she explained. In cat after
cat, the data landed on different points along the same line—all, that is,
except the cheetah's.

"The graph made me realize that the cheetah's inner ear is not
simply different but dramatically different," Grohé recalled. "I had suspected it might be larger, but I was surprised by how extremely large it

was when compared to the twelve other cat species: every main group of wild cat, as well as the highly derived domestic cat." Body mass aside, the balance organs of most cats are between 26 and 36 percent of the inner ear, whereas those of the cheetah take up as much as 44 percent. "Those percentages do not mean that the organ of hearing is comparatively smaller in the cheetah, but that the vestibular apparatus takes up more space in the inner ear," she added. "Relative to body mass, we realized that the cheetah has the largest known vestibular apparatus of any cat by far."

Scientists have long observed that larger balance organs are linked to a fast and agile lifestyle. Even after accounting for body mass, these vestibular spirit levels are decidedly bigger in, say, acrobatic gibbons and galagos than sloths and slow lorises, or flighty gulls than sluggish ducks and geese. The pattern is so widespread across the animal kingdom that scientists have called it "a basic biological phenomenon." The theory is that bigger balance organs are more sensitive to the smallest of movements. "Obviously the skulls we scanned no longer had soft tissue but spaces where the otolith organs and canals once were," Grohé explained. "The larger spaces indicate that there was once more soft tissue, more sensory surface area, and more hair cells. Together, this gives the cheetah an extraordinary vestibular sense." The fluid grace of Sara and her sisters stems from this acute sense of balance.

Grohé turned her attention to the semicircular canals. Using three-dimensional geometric morphometric software, she began to analyze their shape. "I sited digital points along the midline of each of the cheetah's three canals and marked out a total of eighty landmarks throughout." Then she repeated the exercise and assigned the same number of landmarks across every other cat canal system. "The idea is that every point corresponds—so point 1 with point 1 across all canal sets, point 2 with point 2, and so on, until the final eightieth point." By comparing these to one another, Grohé could unmask all variations in canal configuration. "The results showed how profoundly different the morphology of the cheetah's vestibular system is to all the other twelve cat species," she told me. Certain parts of their inner ears were particularly enlarged. "Although its lateral canal is proportionately smaller, its anterior and posterior canals are dramatically elongated." Much of the size increase is found in the two canals that monitor vertical movement: the up-and-down pitch of a nod and the ear-to-shoulder roll of the head. But why should increasing sensitivity to movement in these vertical planes improve the cheetah's balance? This pattern had appeared once before, a few million years ago, in a family far closer to home.

✳

Paleoanthropologists relish a good debate about what it is precisely that makes humans "human." A few qualities distinguish us from our closest living relatives, the other great apes: orangutans, gorillas, common chimpanzees, and bonobos. Big brains are considered our defining trait, as they led to language, technology, and art. As Plato hinted when he called us "featherless bipeds," at some point we also freed ourselves from the floor to stand tall and walk. For long years, the hominin whose very name denotes verticality, *Homo erectus*, was hailed as the first biped. After all, less than 2 million years ago, he marched our ancestors out of Africa. Then, in the 1970s, two discoveries cast aspersions on his title. The first was an Ethiopian find of a small-bodied female dated to 3.2 million years ago. This small-brained australopithecine—named after the expedition's favorite Beatles song, "Lucy in the Sky with Diamonds"—had long and muscular ape-like arms. Lucy also had a broad pelvis with load-bearing thigh bones, suggesting she might have walked. Next came the unearthing of some fossilized footprints at Laetoli in Tanzania. The hominin who had left these tracks had done so when standing on two feet some 3.5 million years ago. Given that these scenarios took place before our brains ballooned, arguably bipedalism began the journey to being human. It remained unclear who had taken the first step—until one scientist realized he could look at the same old evidence in a new way.

Fred Spoor, now a professor in human evolution at London's Natural History Museum, was doing postdoctoral research at Utrecht University in the Netherlands in the 1990s. "There had been some interest in looking at the inner ear of fossils to understand how ancient hominids could perceive sound. This was obviously the cochlea part of the inner ear," he told me, "but I was the first person to concentrate on the other part of fossilized labyrinths." Spoor recognized that the evolution of walking upright must have left its mark on the balance apparatus. He had recently been introduced to an imaging expert at Utrecht University Hospital, Frans Zonneveld. Together, they would pioneer the use of medical CT scanners to look at the same balance organ within ancient hominin skulls that Grohé came to focus on in cheetahs. "We concentrated on one of the smallest and most delicate structures in the head: the semicircular canals," he said. "Most sensory organs don't fossilize well—they are mainly made up of soft tissue like the eyeball—but the canals of the bony inner ear leave a really good imprint." These fossilized balance organs would enable Spoor and Zonneveld to travel back in time and walk alongside our forebears. They scanned an array of past hominin specimens: from *Homo erectus*, via older *Homo* species such as *habilis*, to the more ancient australopithecines like Lucy. They also processed the skulls of thirty-one living primates, including modern humans and all other great apes. By juxtaposing this evidence, they pinpointed the moment in our past when our sense of balance underwent a dramatic shift.

All the inner ears of living creatures held three similarly sized semicircular canals, with one exception: much as Grohé would prove the cheetah is unique among cats, Spoor found we are unique among primates. The size of our canals sets us apart from our closest primate relatives and from all fossil hominins bar one. Spoor explained, "Among the fossil hominids investigated, the earliest species to demonstrate modern human morphology is *Homo erectus*." Despite the conflicting evidence of fossilized skeletons and footprint trails, Spoor inferred that although australopithecines like Lucy could walk on two feet, they still clambered around on four limbs. True bipedalism, not simply walking, but doing so with a hop, skip, and a jump, began in *Homo erectus*. Spoor wrote his thesis and submitted it to the journal *Nature*. He recalled, "The discovery was enormously exciting and what's more, in a sense, still stands."

Pun aside, it is as valid today despite subsequent discoveries of fossils predating Lucy that in all likelihood could walk. His findings piqued Grohé's interest: "Fred Spoor made the inner ear fashionable again, and his studies were enormously relevant to mine. Enlarged anterior and posterior canals were what I was seeing in the modern cheetah." Evolution had emphasized the vertical over the horizontal when early hominins walked on two feet, then again in the far-from-bipedal cheetah. To understand this, science would need to unravel how the semicircular canals work. Inspiration would come from an electrically explosive source.

*

The neuroscientist Brian Day recently retired from running the Whole-Body Sensorimotor Lab at London's University College, where he had dedicated much of his career to unmasking the deceptive simplicity of a leisurely walk in the park. "Humans are superb balancers. Two legs are inherently unstable so every step is a fall forward and sideways toward the lifted foot," he told me. "This makes even the slowest stroll among one of the most daring balancing acts in the animal kingdom." Day became fascinated by the role our semicircular canals play in such feats. "We knew that the canals signal to the brain the mix of head rotations, but nobody had worked out how the brain uses the information to stand and walk." Day learned of some radical studies by the Italian mastermind behind the battery, whose name lives on as the measure of electric potential.

In 1790, Alessandro Volta decided to try out his new invention to investigate Aristotle's five senses on himself. He attached the electrodes to each of his sense organs in turn and flicked the switch. Applying a shock to his eye, he saw light; to his skin, he felt pain; to his tongue, he perceived taste. But when he inserted them into his ears, he could not account fully for the effect. The forty-odd zinc and silver elements shot 30 volts through his inner ear. As well as hearing the sound of boiling matter and an explosion within his head, his world upended and started to spin. Day explained, "The boiling was probably auditory stimulation, though it could have been the sound of actual flesh bubbling, but Volta could not have understood why his world had suddenly started to turn." After all, it would be another four decades before Flourens would uncover how the inner ear affects balance and, even then, in birds. "These feelings were clearly manifestations of vestibular stimulation," added Day. "I began to wonder whether we could use electricity to explore our sense of balance." The thought led to some revolutionary research.

In 2006, Day and colleagues turned to the process of galvanic vestibular stimulation to study how the brain controls walking. By attaching electrodes to the mastoid bones just behind the ears, they could stimulate the inner ear and target its semicircular canals. Simply altering the position of a subject's head created different virtual rotations, fooling their brain into thinking that they were turning about the horizontal or vertical plane. When blindfolded subjects were exposed to virtual horizontal turns as they walked, they subconsciously adjusted their course in the opposite direction. Day and his team were so confident of this effect that they headed to the local botanic garden to put their volunteers to the test. "The results were so strong that we could actually steer blindfolded people by remote control and with such accuracy that they didn't veer off the narrow, winding paths." Day concluded that the horizontal input, which comes from the lateral canals when the head is in its normal upright position, must afford us navigational cues. However, focusing on the vertical aspect had very different ramifications.

"When we adjusted the input that usually comes from the vertical anterior and posterior canals, we actually produced balance disturbances in our subjects," said Day. Flipping the switch to create a virtual vertical rotation made the blindfolded subjects instantly and subconsciously tilt like Mr. MacGregor, almost to the point of keeling over. "We were on hand to stop anyone from injuring themselves," Day assured me. "Generally other senses kicked in so they righted themselves just in time." He tried it on himself. "I hardly felt the virtual input at all, which is not surprising as our vestibular sense is silent, but I quickly sensed my body's response. I was actually swaying and had to take a step to stop myself from falling. It was an odd experience, as if my brain had done something underneath me." Day and his team had identified two different outputs from the semicircular canals that enable us to walk. Information on horizontal head rotations, usually from the lateral canals, allows us to control navigation. Information on vertical rotations, usually from the anterior and posterior canals, affords us the balance to stand upright and step forward. The research provided a causative link between the canals that Spoor had found enlarged and the stability needed for bipedalism, but how they work to keep us vertical goes beyond the inner ear.

"Our vestibular system does not operate in isolation," explained Day. "The brain integrates its input with other incoming sensory information about where the body is in space." Consequently, our eyes play a key role: "Balance is as much about our vision as our vestibular system." Just as our ear is not simply an organ of hearing, our eye is not simply an organ of sight; both are also organs of balance. Furthermore, a crucial aspect of our vision is controlled by our inner ear's semicircular canals. Day explained, "We know that the canals also trigger two bodily reflexes that bypass the need for conscious thought, called the vestibulocollic and vestibulo-ocular reflexes." The first allows unthinking adjustment to the neck musculature to steady the head as the body moves beneath it. The second corrects the muscles in our eye sockets to stabilize the eyeballs as the head moves about them. Both serve to keep our gaze fixed on our surroundings as we move through them. Day elaborated, "The vestibulocollic and vestibulo-ocular reflexes ensure that when we move, the world still appears stable; they turn our eyes into a movie Steadicam." If Spoor showed that large vertical canals accompanied the evolution of true bipedalism in humans, Day explored the reasons. Their increased size hone the reflexes for steady sight and grant us balance as we walk, waltz, and sprint through the world.

＊

Camille Grohé returned to her quadruped. She started to watch slow-motion videos of the cheetah running. "I'd seen plenty before, but this time I really studied them," she told me. After a few viewings, she saw beyond the liquid coordination of powerful backs and legs. "I saw how the animal's head hardly moves at all." On a straight course, despite legs lunging back and forth, the cheetah's head travels at roughly the same height and always parallel to the ground. "Normally a quadruped animal's head moves up and down," she said. "I had never seen such head stability before in carnivores." The penny dropped. "Studies show that agile animals living in three-dimensional habitats—like forests, oceans, or skies—often have bigger lateral canals as it gives them better navigational ability," with finer control over their high-speed leaps, dives, and swoops. "But the cheetah's anterior and posterior canals make it more responsive to the vertical head movements involved in a hunt and increase the sensitivity of two reflex adjustments that stabilize its head and its eyes." The same automatic reflexes that work in us allow the cheetah to lock its gaze on the target. "This means that the cheetah has remarkable visual, as well as postural, stability," said Grohé. Ultimately the enlarged vestibular system helps steady the high-speed pursuit, but the enlarged vertical canals transform the cheetah into a gazelle-seeking missile.

Like Spoor, Grohé then looked to the past. She contacted fossil collections in France that held two important specimens of extinct cats and was soon in possession of two additional virtual skulls from the felid evolutionary family tree. She wanted to compare the cheetah first to *Proailurus lemanensis*, the oldest known cat from around 20 million years ago, and second to its nearest ancestor, *Acinonyx pardinensis*, the legendary giant cheetah, from only 1 million years ago. She extracted the fossils' inner ear dimensions. Like all the other cats apart from the cheetah, *Proailurus* had a comparatively small vestibular apparatus. "I expected this; after all, it is the presumed ancestor of the whole cat family, but I did not expect the giant cheetah data," she told me. Taking body mass into account, its balance organs, including the semicircular canals, were closer in size to those of the tiger or leopard. "They showed none of the modern cheetah's enlargement." This extinct behemoth was twice the weight of today's cheetah, so is unlikely to have run fast, but the findings proved that it also lacked its successor's agility and gaze stability. "The extraordinary hunter we see today in the modern cheetah is clearly the result of late and rapid evolution"—what another paleontologist calls "a recent evolutionary spin-off." Grohé believes the cat's sharpening sense of balance was a catalyst for this transformation: "I think the evolution of the motion-sensitive inner ear opened up the possibility for the cheetah to evolve into the fastest and most agile runner on the planet." So where does this leave patients like Mr. MacGregor whose balance fails them?

✳

Barry Seemungal is a neurologist at Charing Cross and St. Mary's Hospitals in London. "In clinic, I see patients with chronic dizziness, meaning they've been this way not just for days but months, sometimes years." He encounters all sorts of cases. "The definition of dizziness is ambiguous," he said. "To Shakespeare, it meant 'stupid,' but to me it is the feeling of being unbalanced." This might mean being off-kilter like Mr. MacGregor or having a sense of movement—rocking or spinning— when stock-still. Seemungal is interested in how brains receive information from our body to create perceptions of balance. "For example, some patients complain of feeling dizzy only when they see things moving," he said. If you have a problem with your inner ear, the brain starts to pay more attention to the signals coming from the eyes. "That makes sense to a point, but if it places too much reliance [on them], even trivial visual motion can generate false sensations of self-motion." Sometimes the fault can arise elsewhere. Oliver Sacks put Mr. MacGregor's condition down to Parkinson's disease, saying, "One often sees Parkinsonian patients sitting at the most grossly tilted positions, with no awareness that this is the case." He added that the older man's spirit levels were not at fault, but that he had an "inability to use them." Similarly, many patients with perfectly healthy inner ears pass through the doors of Seemungal's clinic. The causes of dizziness are varied and often complex. To understand them better, he turned to a group of people with seemingly supernatural balance.

The whipped throw of a fouetté may be an everyday movement for a ballet dancer, but it leaves the rest of us at best giddy. "Studies had already shown that dancers are resistant to dizziness," said Seemungal. "We hypothesized that they are not born with some innate capacity, but that it is something to do with their training." Ballet dancers are taught certain techniques to counter dizziness. For example, they might fix their gaze to one spot; their head stays still as their body begins to rotate, snapping around at the last moment. This "spotting" works to subdue the inner ear's response to being spun. "Imagine that I am pirouetting for twenty seconds—and trust me, this is not something I normally do," Seemungal explained. "If every half-second I make a head flick, I convert that twenty-second spin into lots of half-second rotations." This prevents the fluid endolymph in the ear from gathering speed and firing up the vestibular system. However, such techniques could only go some way toward explaining a dancer's flair for balance.

Seemungal decided to pit dancers against another equally fit group whose daily routine was less of a whirl. As well as enlisting ballerinas from the Royal Academy of Dance, the London Studio Centre, and the Central School of Ballet, he found women from local rowing clubs. First, the researchers tested how susceptible their new recruits were to feelings of dizziness: each one was strapped into a motorized chair in a darkroom and then spun. "As the chair started to spin to the right, the fluid in their two horizontal semicircular canals was left behind and generated the sensation of dizziness." After thirty seconds, this signal subsided because the speed of the chair's movement did not change. "So although they were still rotating in the dark, they felt as if they were sitting still; from the brain's point of view, they *were* still." After sixty seconds, the chair stopped abruptly. "Although they were now still, their fluid kept moving, and they perceived this violent sensation of spinning to the left." The subjects were asked to turn a little wheel to indicate the direction and speed of their perception of this continued spin, while electrodes attached to their eyes measured their physiological response. "The first thing we found was that, unsurprisingly, the dancers' duration and intensity of dizziness following being spun around was much less than that of the rowers, both physiologically and perceptually," explained Seemungal. The ballerinas were indeed more pirouette-proof.

Each volunteer was given a brain scan. Some fifty scans later, the researchers were faced with a very clear pattern. It exposed a disparity in an area of the brain called the cerebellum that receives information from the inner ear. Seemungal clarified, "We found the difference in the particular part of the cerebellum that processes balance, the gray matter." One might think that this area would be more developed, like a muscle. However, the gray matter was not only reduced significantly in the ballerinas; it was also smallest in those with the most experience, suggesting it had shrunk with practice. "The cerebellum processes signals from the balance apparatus and then sends it to cortical areas of the brain linked to perception," explained Seemungal. "It acts like a gate, and in the dancers, this gate was reducing the flow of signals." In Chapter 4 on the star-nosed mole, we learned that the brain of a blind braille reader adapts to magnify the sense of touch. That of a dancer also changes, but to suppress information coming from the inner ear and downplay the sense of balance. "This happens because dancers are over-practiced animals whose movements are always correct. They don't need feedback from the vestibular system telling them where they are; they already know exactly where their body is in space." That their daily routine serves to reduce their reliance on their organs of balance has implications that promise much for Seemungal's patients.

The study stresses the fundamental role of the brain in our sense of equilibrium. "The key to understanding dizziness is to realize it is not an ear problem but a brain problem. In fact, the brain is the most important balance organ in the body." Seemungal's words echo those of Paul Bach-y-Rita: that we do not see with our eyes, we see with our brain, and so—as well as not hearing with our cochleae, feeling with our skin, tasting with our tongue, smelling with our nose—we do not steady ourselves with our otolith organs or semicircular canals. Vitally, whereas the hardware of our inner ears may be fixed and immutable, the software of our brain can be rewired. Such neuroplasticity gives researchers like Seemungal something to work with. Counterintuitively, he plans to treat people with chronic dizziness by getting them up on the dance floor. "It could be ballet. It could be break dancing," he said. "They don't need to be doing complicated dance moves with any degree of proficiency; they just need to be doing exercises that move their head." He hopes that, as with the ballet dancers, this will mold his patients' cerebella to deemphasize their vestibular response and banish their giddiness.

Oliver Sacks's patient took a different tack, using his skills as a craftsman to solve his imbalance. On being told that the spirit levels in his head no longer worked, Mr. MacGregor wondered, "Why couldn't I use levels outside my head—levels I could see, I could use with my eyes." With the help of the clinic's optometrist and workshop, he constructed a pair of what Sacks described as "Heath Robinsonish" spirit spectacles. The prototype glasses bore a clip extending out from the bridge, holding miniature horizontal levels that Mr. MacGregor could eyeball as he walked. Initially it took a great deal of effort to go about his day-to-day life, continually focusing on the levels and straightening himself accordingly. However, as Sacks explained, "Over the ensuing weeks, it got easier and easier; keeping an eye on his 'instruments' became unconscious, like keeping an eye on the instrument panel of one's car while being free to think, chat." Despite his Parkinson's disease, Mr. MacGregor's central organ of balance had adapted to the task at hand, enabling him to once more walk tall.

*

We tend to notice our sense of balance only when things go wrong—when we end up, like Mr. MacGregor, at a precarious angle, like one of Barry Seemungal's patients feeling disorientated, or, like Alessandro Volta, sprawled on the floor. It works tirelessly and without conscious perception. It is one of our many secret senses. "We still know so little about the vestibular sense," lamented Brian Day. "It is involved in many aspects of movement, not just balance but navigation, spatial orientation, even eye control, and recent studies suggest it also helps us predict the movement of objects beyond ourselves." We know that

100,000 years ago, or thereabouts, it enabled the evolution of the most sure- and fleet-footed animal on earth. Millions of years earlier still, it raised our ancestors from the ground and set them on the path to becoming human. Yet science is only beginning to expose the many other moments, through deep time and in our everyday, that are made possible by this unsung sense.

10

The Trashline Orbweaver
and Our Sense of Time

A SPIDER CROUCHES AMONG THE DEAD, MOTIONLESS AT the center of her cobweb. She hides within a woven column of eviscerated abdomens, decapitated thoraxes, and stray legs. This graveyard of flies caught, killed, carefully disassembled, and then reassembled is how she conceals herself. The trashline orbweaver (*Cyclosa turbinata*) is a virtuoso of disguise, and her web is her camouflage. Its filaments radiate outward like a gossamer extension of the spider's body. Each of her eight legs rests on separate silken spokes. Sometimes she plucks or bounces the strands to sense their tension and locate any damage; other times she sits and waits for the tiniest of incoming tremors. Scientists have shown that orbweavers can detect movements a mere thousandth the width of a human hair. She then distinguishes vibrations made by wind from approaching male suitors or those with more nutritional promise. A fly ensnared within her web's sticky spirals will struggle to free itself. The spider "listens" to its ever-more jerky motions, then darts with disarming speed, tiptoeing along the web's nonsticky radial byways. If the prey is large, she casts a silken throw from her spinnerets, tightly cocoons it, and sinks her fangs through its rigid exoskeleton. Her venom glands contract and paralyze. The spider then sucks the fly dry and weaves its body husk into her gruesome trashline. The orbweaver's web enables her to feel a wider world, but its creation comes under the influence of another rarely recognized and secret sense.

Few students dare to pass through the door into Thomas Jones's office at East Tennessee State University. It bears a picture of a fearsome and fanged invertebrate, emblazoned with the words "Spider Lab." Transparent plastic boxes line the walls, containing a bewildering array of arachnids. "This is Rosie, our rose-haired tarantula," said Jones, pointing to a specimen nearly as big as his hand, but in appearance more soft toy than spider. Then he motioned toward another box. A small black shadow hides in the corner, its belly displaying the hallmark red hourglass. "Meet our most venomous spider, *Latrodectus mactans*, the infamous black widow." Reaching up and pulling out a third box, Jones lifted the lid; a hairless, pointed, and articulated leg emerged, followed by a large gold-and-black abdomen. Jones coaxed it onto his hand. "This is a female *Nephila*, a golden orbweaver." She almost filled his palm. "Of the thousands of orb-weaving species in North America, she is our largest. She looks impressive, but she's harmless."

The entire back of the Spider Lab is devoted to one of Jones's smallest orbweavers. Strung between lines of wooden frames are a series of webs with the distinctive trashline, each hiding a body mere millimeters across. "I'm interested in the personalities of spiders," Jones said, "and the tiny trashline orbweaver has a big personality." She is never without her trashline—if she weaves a web elsewhere, she lugs it along—so when Jones sent his students into the woods on campus to collect these specimens, he gave strict instructions not to omit the silken luggage. "Each trashline contains the corpses of all the flies that the spider has ever eaten," he told me. "It's quite repulsive—perhaps macabre when you think about it." Back in the lab, they noticed something peculiar. "The spiders are diurnal. We watched them hunt through the day and saw their webs slowly accumulate damage, but we never saw them make repairs." In the wild, along with flailing flies, webs are marred by pollen, leaves, twigs, even dew and humidity. "Mysteriously, every morning the webs appeared as good as new." Jones began to wonder whether this spider was also up at night. "Nobody had considered how time of day affects spiders' behavior, but just upstairs from my lab is a world expert on biological rhythms."

Darrell Moore has spent the past four decades—most of his career—studying the honeybee. "I didn't think I would start work on another species," he admitted, "but when TJ knocked on my door and was so excited, I couldn't resist." Moore analyzes behavioral rhythms with an experimental device called a locomotor activity monitor. It relies on a compact array of crisscrossing, constant beams of infrared light to measure motion within a test tube. So when a subject—bee or spider—moves, it cannot help but break a beam. A computer records every minute of every day for several days, plotting the creature's activity against time. These graphs are called actograms. "Essentially the locomotor activity monitor acts like our spy, saving us from having to watch the whole time," explained Moore. "We simply leave our subjects to do their thing, undisturbed, and return for the results."

Moore and Jones decided to investigate twelve trashline orbweavers over five days. First, they had to ensure that the spiders were offered suitable sustenance. Jones told me, "I decided that one fruit fly and two drops of water per test tube were more than enough for one spider." Next they stoppered the tubes loosely so the occupants could breathe and slotted them into a rack within an environmentally controlled room. "I chose a temperature and humidity to mimic the spider's natural habitat, and a light–dark cycle with twelve hours each way, so sunrise at 8:00 a.m. and sunset at 8:00 p.m.," recalled Moore. On the evening of the fifth day, Moore consulted the twelve actograms. To his trained eye, a pattern was immediately apparent: all the spiders were on much the same routine—active during the day as expected and even more so at night. Moore said, "The spiders were showing peaks of activity for two hours from 4:00 a.m., then became inactive again before the sun came up." Jones explained, "We realized they weren't just repairing their webs under cover of darkness; they were rebuilding them from scratch every night." These so-called diurnal spiders were waking in the dead of night, before sunrise. They were not responding to light so much as preempting it, and they were doing so with the regularity of clockwork.

*

In the balmy Languedoc summer of 1729, the astronomer Jean Jacques d'Ortous de Mairan, a man whom Voltaire would laud as one of the five most remarkable scholars of the eighteenth century, turned away from his usual preoccupations with planetary phenomena, solar eclipses, and the aurora borealis to a common or garden plant: a flowering mimosa from the pea family, known colloquially as a touch-me-not. Perhaps he had encountered the mimosa on a walk through the grounds of his home in Béziers or in one of Europe's oldest botanical gardens at nearby Montpelier. Either way, de Mairan observed its filigreed fronds shrink from his touch and became intrigued by how they moved throughout the day when left to their own devices. He noted the foliage rise with the sun, unfold to reveal rows of smaller leaves, refold, and fall with the sun. He then watched the cycle repeat itself as sure as day follows night.

A friend presented de Mairan's findings as an *observation botanique* to the Royal Academy of Sciences in Paris: "We know that the Sensitive-plant is a heliotrope, that is to say that its branches and leaves always turn in the direction of where there is most light." Curious to discover what would happen if the mimosa was separated from the very thing it seemed to track, he took a potted specimen and closeted it in a cupboard. Over the next few days, he cracked open the door, careful not to admit any light, and watched to see what, if anything, was happening. "The experiment was performed toward the end of summer, and repeated several times. . . . It appears to sense the action of the Sun and daylight, even when not exposed to them." Like the orbweaver, the plant seemed able to anticipate sunlight. In fact, this simple experiment demonstrated an ability to sense something other than, though intimately connected to, light. De Mairan would never realize the full import of his work, but he was the first to show that this species of mimosa can sense time.

Life began on this planet some 3.8 billion years ago. Since then, the earth's rotation has slowed so that today, one revolution takes just under twenty-four hours. "Twenty-three hours, fifty-six minutes, and four seconds to be precise," said the neuroscientist Russell Foster. The period when we face the sun, bathed in light, we call day, and as the earth turns, darkness and night descend. Around the planet, life unfolds in predictable sequences every twenty-four(ish) hours. Even before the first rays of sun filter through an ancient beech forest, a mouse stirs. As the sun rises, birdsong swells. Flowers blossom and honeybees arrive, their visits repeating through the day with scheduled precision. In the oceans, plankton soar up the water column, then sink with the sun. Just before twilight, the eyes of fish on a coral reef ready themselves with a switch from day to night optics.

Dusk descends on a Madagascan jungle, and the klaxon calls of ring-tailed lemurs cease. They fall asleep on one another's shoulders on high branches, just as their distant relative, the shy aye-aye, opens its eyes on its "day." Scientists have explored these natural quotidian rhythms. Like de Mairan, they have deprived creatures of light and watched certain behaviors persist, proving that the trashline orbweaver and the mimosa are not alone in being able to know the time of day. Nearly all plants and all animals, as well as fungi, algae, and even some bacteria, march to the beat of an internal timekeeper. This is a clock that is so reliable, so accurate, that some say it possesses the hallmarks of one crafted by human hand. Yet whereas a mechanical clock marks human notions of hours, minutes, and seconds, a body clock gauges natural time and the passing of a solar day. The questions of whether such clocks tick silently within us—whether in the absence of sundials, hourglasses, wristwatches, mobile phones, or daylight we too can unknowingly sense time—inspired one of science's most daring self-experiments.

*

On July 16, 1962, the speleologist Michel Siffre set off into Scarasson Cavern, deep beneath the French Alps. After descending its vertiginous, 40-meter (130-foot) shaft and pitching camp, he bid his colleagues *au revoir* and watched as their line of head torches bobbed away, swallowed by the blackness. "I decided to live like an animal, without a watch, in the dark, without knowing the time." His book *Beyond Time* recounts what happened over the following two months. "The cave was completely dark, with just a light bulb." Cut loose from a predetermined day and night, he planned to listen to his body. He would eat only when hungry, drink only when thirsty. He would switch off the light when tired and fall asleep; then switch it back on and resurface when rested. His sole link to the outside world was a phone line rigged up to the mouth of the cave, continually monitored. He and his guardians had agreed on a simple protocol. "I would call them when I woke up, when I ate, and just before I went to sleep," he explained. "My team didn't have the right to call me, so that I wouldn't have any idea about what time it was on the outside." After Siffre's first sleep, accompanied by the inevitable loss of consciousness, the experiment got underway. The team on the surface kept vigil, silently recording the hours of their subject's waking, eating, and sleeping, and by day two, they found that Siffre was already two hours out of kilter.

As days merged into nights, Siffre realized that although his food stores stayed fresh for longer than hoped, the conditions proved challenging. "I had bad equipment," Siffre recounted. When he slept, the only thing separating him from the icy floor was a mattress made from thick sponge, which soaked up the damp and chilled his bones. The atmosphere was permanently dank, and his clothes failed to dry out overnight, making getting dressed the next morning particularly unpleasant. "My body temperature got as low as 34°C [93.2°F]," he explained. "My feet were always wet." He became increasingly lonely. For company, he turned to another cave dweller, albeit one that would not have looked out of place in Thomas Jones's laboratory. Siffre put the spider in a box; he fed it, watered it, and watched it watch him.

As he counted down his stay, he received an unexpected phone call. His ground team announced that his time was up. Siffre was convinced there were several more days to go, but they insisted that two months had passed. He was in such a weakened state that he had to be winched to the surface, and, blacking out, he was swiftly helicoptered off the mountain. But the ordeal would prove worthwhile. Siffre had written in his diary, "Time no longer has any meaning for me. I am detached from it, I live outside time." Yet despite this perception and being wrong about the length of time he had spent in the cave, the data gathered revealed a different story. It transpired that throughout, Siffre had adhered unwittingly to a regular wake and sleep cycle. "Without knowing it, I had created the field of human chronobiology," he said. "My experiment showed that humans, like [other animals], have a body clock." Other cave isolation studies followed, each more audacious than the last. In 1965, Josie Laures lasted 88 days, Antoine Senni 126, and in 1972, Siffre once again broke records, this time in the depths of Midnight Cave, Texas, with 205 days—six months. Today, he rails against new rules that forbid such exploits: "When we first did them, I was young, and we took all the risk [but now] the experiments in the caves are finished." However, he and his pioneering colleagues were proof that, like his pet spider, humans do indeed have a body clock: one that marks the passing of time without conscious sensation. The answers to how it works, and many other questions, would be found beneath the rolling hills of Bavaria, near the Heilige Berg, the Holy Mountain, and the Benedictine monastery of Kloster Andechs.

Jürgen Aschoff, a key scientist in this emerging field of chronobiology, learned of Siffre's exploits and wondered how to replicate them with scientific rigor. A promotion to the department director at the new Max Planck Institute for Behavioral Physiology in Andechs and the discovery of a nearby World War II bunker presented the means. With funding from NASA, he transformed the decommissioned, dilapidated bomb shelter into an isolation unit. Made lightproof, soundproof, and vibration proof and encased in a copper cage, even shielded from daily variations in the earth's electromagnetic field, it was more cut off from the world than Siffre's cave. Yet it also boasted comparatively luxurious living quarters: a sitting room that doubled as a bedroom, along with a shower and a small kitchen. The only entrance was via a corridor with thick doors at either end. As the subject walked from the exterior world toward their new home, an outer door was double-locked behind them. A magnetic catch ensured this had to be closed before an inner door would open, so once inside, the subject would see neither daylight nor another person throughout the stay. Watches, clocks, phones, televisions, and radios were all left behind. "The only way in which the subject can communicate with the outside world is by sending and receiving letters," explained Aschoff. These were passed through the corridor, along with food and an occasional bottle of the monastery's beer. The subjects were told to turn the lights on when they woke, switch them off when they went to bed, and to cook and eat three meals each day. "We ask him to lead a 'regular' life," Aschoff added; "otherwise he is allowed to do what he wishes." Yet these men and women would be monitored meticulously. Their temperature was measured, their urine analyzed, and, like Moore's spider locomotor activity monitor, electrical contacts embedded within the bunker floor would record their every step to produce human actograms.

Aschoff was first through the double-locked door. He recalled his experience in a paper he wrote for the journal *Science*: "After a great curiosity about 'true' time during the first two days of bunker life, I lost all interest in this matter and felt perfectly comfortable to live 'timeless.'" He continued, "In the 'mornings,' I had difficulty deciding whether I had slept long enough. On day eight . . . shortly after breakfast I wrote in my diary: 'Something must be wrong.'" The scientists outside the bunker could see that Aschoff had only had three hours of sleep, but something then prompted him to return to bed, sleep, and restart his day at a more seemly hour. Much like Siffre before him, Aschoff was developing a remarkably stable rhythm. "When I was released on day ten," Aschoff recalled, "I was therefore highly surprised to be told that my last waking-up time was 3 p.m." Again like Siffre, he had slowly drifted against real time. The actograms revealed that his body clock, though regular, had been cycling around twenty-five hours, adding almost an hour to every day. This shift, though small, could cause a surprising level of disruption. For subjects who spent longer periods in the bunker, day twelve might see 6:00 a.m. slip to 6:00 p.m., turning day into night. Day twenty-four would likely bring them back in sync with time, but having lived through only twenty-three, slightly elongated days. The actograms of Aschoff and his test subjects solved the paradox of how Siffre's body clock had been so regular yet also inaccurate. This would have been the inevitable consequence of two months' free-running at approximately the same rate as Aschoff's.

Over the next twenty-four years, more than 400 further trials took place; only four volunteers asked for early release, and the rest produced a vast body of evidence. The body clock of one student, who had entered the bunker to cram for exams, spun out of control with cycles as long as thirty-three hours; on his release, and to his dismay, exam day was more imminent than expected. Otherwise, body clocks behaved much like Aschoff's, keeping days that were, on average, longer than twenty-four hours but shorter than twenty-five. The bunker proved, incontrovertibly, that the human body clock is circadian, from the Latin *circa diem*, "about a day." It matches the natural day closely but not completely. The Andechs team never ascertained how we sense the passing of a day. The answer had been staring them in the face all along.

✳

On August 31, 1999, British Army sergeant Mark Threadgold suffered a near-fatal head injury. His brain sustained so much damage that he was transferred among three hospitals as different teams of neurosurgeons battled to save what functionality they could. During this period, Threadgold drifted in and out of consciousness. He vaguely recalls being told he had lost his senses of smell and taste, also that he was blind. He refused to believe it. "I had wet pads taped over my eyes, and I remember explaining to a nurse that my eyes were fine, that I could see perfectly," he told me. "To prove my point, I told her she had long dark hair. I must have been totally spaced out." A month later, with wounds healed and surfacing from the morphine haze, the gravity of his situation hit home. "I didn't want to eat; I couldn't smell or taste. I didn't want to get up. What was the point? Although I could walk around the wards by now, I just lay on my bed. The environment was strange, everyone was a stranger, and I couldn't see anything at all."

Threadgold's eyes remain intact, but his optic nerve had been destroyed, so no information can reach his brain. He calls his sight loss "black blindness." His days are like nights, the darkness absolute and unrelenting. Eventually he was sent to a specialist rehabilitation center where he was given a clock that speaks the time. "At least the time was one thing I didn't have to ask someone for." It soon alerted him to something strange. "One evening after I'd got the clock, I'd gone to bed around 9 p.m. The next thing I knew, I had woken up and the clock said it was 11:30." Surprised at how long he had slept in, he quickly shaved, showered, and went downstairs, only to find the building quiet. "Weirdly quiet," he told me. Slowly, it dawned on him that there was no one around. "I was terrified. Again, I asked my clock the time. It was actually 11:30 p.m. Everyone had only just gone to bed."

Threadgold realized that time was proving unreliable. The slippages worsened when he left the rehabilitation center to live on his own. A nap after lunch would last hours. He would wake up late in the evening, believing it to be morning. "Within my first week, I found my days becoming inverted. I was up, unable to sleep through the night, exhausted, and unable to stay awake during the day. Then another week, I'd slip back into the normal rhythm." As it had for Siffre in his underground cave and Aschoff in his bunker, Threadgold's body clock had warped and elongated. He was drifting in and out of time, losing entire days. The disruption was unrelenting, and the nocturnal spells isolated him from the rest of the world. He spiraled into depression. "It was soul-destroying," he explained. "That was when I hit rock bottom." Doctors neither recognized nor understood the nature of Threadgold's challenge. At least some other blind veterans were able to give him advice. "They showed me that the only way to beat this is to keep to a strict and meticulous routine when I wake, sleep, but also eat and exercise," he said. "I can't have a day off. I have to live every hour of every day by this clock, even when it doesn't feel right." With military precision, he has replaced his body clock with the talking clock, listening more carefully to whether it says *ante* or *post meridiem*. Threadgold has since been diagnosed with an uncommon condition; his visual blindness has somehow rendered him "time blind."

Russell Foster was one of the few scientists interested in Threadgold and similar cases. He wondered whether they might illuminate how we all sense time. "There is no use having a clock if you can't set it to the time of day," he told me. The Aschoff bunker had gone on to show that light acts like a daily adjustment to the winder on a mechanical watch, setting our body clock to the correct time. The way in which this light is sensed, and where, remained subjects of intense speculation. As discussed in the previous chapter, it took until the early nineteenth century for it to be discovered that the ear concealed a sensory ability beyond the obvious: a secret sense for balance. Foster wondered whether the same could be true of the eye. "Patients like Mark Threadgold made me realize the importance of the eye, but no one was asking how it enables our sense of time." Ophthalmologists believed that the human eye was the most understood organ in the history of anatomy and neuroscience. How its two light receptors—the cones and the rods—give us color and dark vision had been mapped comprehensively. So the idea they might also serve another purpose was considered preposterous. "But you see, I wasn't a human vision scientist so perhaps my naiveté helped." At the time, Foster described himself as a photobiologist whose inspiration was the natural world. Today, he runs the Nuffield Laboratory of Ophthalmology in Oxford.

"My first foray into the weird world of photoreceptors was with the lamprey," he told me. After reading all about this ancient, eel-like fish, he started research on amphibians, then moved to birds. Finally, his attention focused on mammals, specifically the mouse. He explained, "There are all sorts of lab mice strains, each with different eye mutations, which are used to investigate blindness." He hoped this tiny mammal would shed light on the human condition of time blindness. "First, I studied a nearly totally blind mouse known as the retinally degenerate or rd/rd mouse." This mouse's eyes are devoid of rods and have very few cones. Foster recorded their actograms while they turned a running wheel. When the results were juxtaposed against those of regular mice, he saw they kept the same schedule. Even without rods, the mutant mice were able to synchronize their body clocks to the natural day. Then he ran the experiments on a cone-less, or cl, mouse. The outcome was the same. To exclude the possibility of any undefined collusion between rods and cones—perhaps one taking over when the other failed—he would need a strain of mouse that had neither, something vision scientists had never had call for before. "Rather than wait for the chance discovery of a mutant mammal, we developed a new mouse strain," he said. "We called it—affectionately if unoriginally—the rd/rd cl mouse." Foster repeated the experiments on this new mouse and compared its actograms against those of regular mice. "It was goosebump time. The actograms were indistinguishable. This mouse with profound visual blindness was not time blind." Before drawing conclusions, he returned to the experiment, but now completely covered their eyes. The blindfolded mice lost all sense of time. The results were clear: "There had to be some other light sensor in the eye that enabled it to stick to a normal circadian rhythm." Foster proposed that the retinas of mice, humans, indeed all mammals have rods, cones, and some unknown third photoreceptor that grants the sense of time.

The reception from the academic community was far from welcoming. "I announced this idea at a scientific conference," Foster recalled. "I was told 'no way' and 'get real.' I was called a fool, an idiot, and one chap stood up, stared me in the eye, then swore at me before walking out. The hostility shocked me." Even Foster's colleagues told him that surely there must be another explanation for how his blind mice sensed time. Ron Douglas, the finder of the four-eyed spookfish, would embark on what he calls "a brief but successful collaboration" with Foster. He said of that time, "I am ashamed to say that when Russell started to apply for grants to fund his work, I was not always very helpful. I was one of the many vision scientists who found it hard to believe there was an as-yet-undescribed photoreceptor in the retina." The establishment was almost unanimous in its refusal to accept the possibility that some 150 years of intense scientific scrutiny of the human eye had missed an entire class of photoreceptor. Foster was downcast but determined to find incontrovertible proof. "I had trespassed across disciplinary fields beyond my own, so my science had to be like Caesar's wife—beyond reproach." Little did he realize that this desire would define the next two decades of research, engaging the help of laboratories around the world.

At around the turn of the new millennium, Foster's team published a microscopic image of a mouse retina, stained to highlight a fine cobweb within the network of ganglion cells; these overlie the rods and cones, their projections gathering together to form the optic nerve. "Importantly, this ganglion cell network is so sparse and diffuse that it does not screen the visual photoreceptors from the light," he told me. Of these cells, one or two in every hundred had absorbed the stain. "These few cells had to be the circadian light receptors." Finally, Foster and his research team had found the eye's third photoreceptor. Its photosensitivity is based on an entirely new opsin photopigment. It is neither the rhodopsin of rods nor the photopsins of cones. It is a pigment called melanopsin that had only recently been uncovered in frog skin. To silence any remaining skeptics, the scientists measured these cells actually responding to light. With a final flourish, they proved that mice born without this opsin are as blind to time as Threadgold.

Deciding on a name for this clock-winding cell has proven almost as complicated as finding it. "We started with 'endogenously light-sensitive retinal ganglion cell,'" said Foster. "Or 'intrinsically photosensitive melanopsin retinal ganglion cell,'" added Douglas. "Now we've settled on 'photosensitive retinal ganglion cell,' pRGC for short." None trips off the tongue like *rod* or *cone*, but, name aside, Foster's theories of a novel sensor at work within us have been proven definitively and accepted. The scientific renegade has been recast as revolutionary. Today, Douglas teases Foster, saying it was his early opposition that galvanized action and ensured success. He conceded, "Despite my lack of help, Russell's work did get funded, and that's lucky for all of us. His discovery of our eye's third photoreceptor is arguably the biggest advance in visual science in the past fifty years." Thanks to Foster, we know now that the eye is not solely an organ of sight that provides our conscious sense of space and plays a vital role in balance. It is also a circadian organ, endowing us with our subconscious and secret sense of time. Photons travel through our pupil to the retina. Here, they hit either rods, cones, or Foster's time-sensing cells. Information from all three travels in unison through the optic nerve and then splits in the brain at the optic chiasm. Output from the rods and cones ends up in the visual cortex, ultimately creating perceptions of the scene before us with chiaroscuro, color, and depth. Signals from the time sensors continue deep within our brain to the base of the hypothalamus. Their destination: a paired clump of cells known as the suprachiasmatic nuclei.

In 2017, a Nobel Prize was awarded to a group of scientists who showed how the body clock ticks in the cells of a fruit fly. "Remarkably, the molecular clock of the fly is very similar in all animals, including us," Foster explained. "We have learned that virtually every cell of our body can generate a circadian rhythm. There are liver clocks, muscle clocks, pancreas clocks, adipose tissue clocks, and clocks of some sort in every organ and tissue examined to date." But those in the suprachiasmatic nuclei rule the rest, keeping them ticking together in time. Foster likened this master clock to the conductor of an orchestra, warning, "If you shoot the conductor, the musicians will keep playing, but they'll drift out of time with one another and the symphony becomes a cacophony." Time blindness leaves people like Mark Threadgold fighting against such desynchrony on a daily basis. Even with a fully functioning retina, without an optic nerve, it is not possible for the responses of its rods, cones, and time cells to reach his brain. Nonetheless, the research he inspired can be applied to many. "It has led to a major reappraisal of all blindness," said Foster. "Some people can be blind, but, unlike Mark, still have a working sense of time. Now we know to insist they spend time in the light they cannot see." Only this will stop their master clocks from losing synchrony with the turn of day and night, thereby losing command of all the other clocks in their body. The consequences of such disruption, through time blindness or regular blindness, extend beyond slippages in sleep and wakefulness. Science is starting to show they may exact an even greater price.

According to the eminent chronobiologist Colin Pittendrigh, "A rose is not necessarily and unqualifiedly a rose; that is to say, it is a very different biochemical system at noon and at midnight." The same is true of spider, sloth, or human. Our master clock choreographs all the reactions that take place in our body; it ensures that they happen in a regular and timely order. We are creatures of habit in strange and surprising ways. "There have been hundreds of studies showing that a broad range of activities, both physical and cognitive, are affected by the time of day," Foster told me. His list is expansive. We feel more tooth pain in the evening. Labor pains likely begin at night, but babies born naturally tend to arrive in the morning. Proofreading and sprint swimming are best done in the evening. "Our ability to perform mathematical functions or other intellectual tasks between 4:00 a.m. and 6:00 a.m. is worse than if we had consumed several shots of whisky." Even though there is far less traffic, road accidents peak at 3:00 a.m. "It's no coincidence that accidents such as Chernobyl and the *Exxon Valdez* occurred on a night shift."

Working through the night, as well as traveling across international time zones, exposes the stark consequences of defying our biological rhythms. "Shift work and jet lag uncouple our master clock from the natural day and from our cellular clocks, so our normal circadian system falls apart," explained Foster. "Metaphorically speaking, your stomach ends up over Peking, your liver is somewhere near Delhi, while your heart is still in San Francisco." Studies into desynchrony illustrate the many ways in which it can affect the body beyond simple tiredness. The side effects read like a drug disclaimer: impaired memory and concentration, as well as cognitive and motor performance; increased risk of irritability, depression, and risk taking; reduced immunity and endocrine function. The relentless desynchrony enforced on shift workers is potentially chronic. Recently, the World Health Organization classified shift work as a risk factor for cancer, but research has also linked it to type 2 diabetes and cardiovascular disease. Circadian researchers are still amassing data but worry that time-blind individuals are as susceptible to such deadly diseases. Foster is now working on drugs that will make the suprachiasmatic nuclei act as if they have been exposed to light. "If I can use these drugs to give back a sense of time to those who have lost their sense of sight, I'll retire happy," said Foster. With such high costs to desynchrony, no wonder our body clocks impose a circadian rhythm. Scientists believed this was the case for all creatures until recently, when a wild deviation was found in the woods.

*

Darrell Moore had decided to let his experiment on the trashline orbweavers run. Day five turned into day six, the lights in the environmentally controlled room were left off, and the subjects were kept in constant darkness under the watchful gaze of the locomotor activity monitor. On the tenth day, Moore looked again at their actograms. Sure enough, despite the absence of light, the spiders' body clocks had kept their behavior cycling in a regular routine. However, he could also see they were freewheeling and clocking "days" at a profoundly odd rate. "*Cyclosa*'s rhythm was far from a twenty-four-hour *circa diem* like ours," he told me. "It was really, really short. In truth, so short that I didn't tell TJ at first because I didn't believe it." Instead, he reran the tests; then again on a different group of trashline orbweavers. Weeks passed, but the outcomes did not change. "There was no denying the results this time," said Moore. "*Cyclosa*'s body clock was free-running at eighteen and a half hours." It was time to tell Thomas Jones.

"I recall perfectly the morning that Darrell came to knock on my door. When he walked in, I could tell something was up." It dawned on them that they had discovered the shortest body clock known in nature. Jones added, "The result made my mind whir. What was going on, and why?"

"We thought finding *Cyclosa*'s close relatives might shed some light," Jones told me. First on their list were the two spiders nearest to the trashline orbweaver on its evolutionary family tree. *Allocyclosa bifurca* and *Gasteracantha cancriformis* do not have common names. "We tried crowdsourcing, asking our students to help with the hunt." The spiders are identified easily by unusual abdomens; *Allocyclosa*'s is forked, whereas *Gasteracantha*'s bears six sharp spines. Soon the students returned from the nearby scrublands with sufficient numbers for the locomotor activity monitor. Moore ran ten of each species for ten days in darkness. "I remember going to collect the actograms," he recalled. "Once again I couldn't believe the results. It turns out *Cyclosa* is not unique: both of these species also have oddly short clocks." *Gasteracantha* displayed a circadian rhythm of nineteen hours, while *Allocyclosa*'s was the shortest yet, at just over seventeen hours. Jones said, "We were struggling to make sense of it all. Because these three spiders are closely related, we could at least suggest that one evolutionary event had potentially given rise to all these short clocks."

The trashline orbweaver's next closest kin is a species called *Acanthepeira stellata*, the star-bellied spider, for its distinctive spiked abdomen. This small creature proved elusive. Days passed, then weeks. The search parties kept returning empty-handed. "We ended up renaming *Acanthe-*

peira the 'unicorn spider,'" said Jones. "I even put up 'WANTED!' posters in my arachnology class." One month later, their luck changed, and the mythical creature was tracked down in a thicket nearly 50 kilometers (30 miles) from their laboratory doorstep. "We found that *Acanthepeira* doesn't have a short clock at all," Jones told me. "Far from it. It was the last thing we were expecting. It has a long clock, a really long clock." *Acanthepeira's* body clock turns at just over twenty-eight hours. Moore recalled, "When we got these data, we held a special session in TJ's class to inform the students who had found all these spiders. There were gasps followed by silence. No one had ever heard of a body clock this long before."

The trashline orbweaver and its wider family boast by far the shortest and longest body clocks in nature. The only creatures that come close are mutants, generated by scientists in laboratories for the purpose of studying circadian rhythms. Moore explained, "Before *Cyclosa*, the shortest clocks that we knew of were mutant hamsters with twenty-hour clocks and mutant fruit flies with nineteen-hour clocks. We also knew of other mutant fruit flies with clocks that were as long as our unicorn's." It was always assumed that these lab creations would not last in the wild. "Time and time again, science has shown animals with circadian clocks that differ from the twenty-four-hour solar day have trouble surviving," said Jones. At that point, both scientists piped up, in perfect synchrony, "So, theoretically, none of these spiders should exist." Yet there they were, their webs strung across the wooden frames lined up in Jones's lab. In the dark hours of the morning, each trashline orbweaver stirred to reweave its web—first structural spokes, then sticky spirals—before hunkering down, motionless, in its fly graveyard as dawn broke. Every one played out the same routine every day, oblivious to one another and to the consternation they had caused.

Sunlight resets the spiders' body clocks to the solar day, so the extreme short and long cycles do not occur in the wild. "But think about it," Jones said. "That means *Cyclosa*, with a body clock of eighteen and a half hours, has to phase shift by more than five hours to stay in sync." The scientists liken it to snapping out of five-hour jet lag. Moore elaborated, "It would be like traveling west from London to New York and readjusting instantly to being five hours behind Greenwich Mean Time." The unicorn has to phase-shift four hours in the other direction. "That would be like traveling east from New York to London," explained Moore. "And remember, we're not just talking of the odd occasion." Human bodies struggle to adjust to a single hour shift each day. These creatures face time phases that would throw us into desynchrony and reset themselves with every single dawn.

The spider clocks could hardly be more different from our own. Moore continued, "When I tell other circadian rhythm researchers about all these findings, their first response is 'What?' And their second response? Well, let's just say you couldn't publish it!" Other laboratories have joined the research effort. The scientists hope to discover what lies behind the spiders' circadian versatility and what benefit their short and long clocks confer, if any. "In experiments we've seen some spiders start with a short cycle and stay short, some start long and stay long, but others start at twenty-three hours, then turn to twenty-five hours, then switch back again," said Moore. "So maybe spiders possess both short and long clocks, and we're seeing the interplay between them." The questions are fast outnumbering the answers.

✳

In the 1930s, the French biophysicist Lecomte du Noüy wrote, "Western man makes clocks with smaller and smaller divisions until he can now measure a millionth of a second. He assumes that the measurement of a fraction of a second represents an absolute measure of some strictly objective reality." In 2017, a paper appeared in *Nature Physics* that fine-sliced time into one-trillionth of a billionth of a second: a zeptosecond. No other creature tells the time as we do. We alone turn to watches that circumscribe hours, minutes, and seconds. We alone devise tools to clock zeptoseconds. Yet these measurements are our inventions. The trashline orbweaver is a talisman for natural time. The radical discovery of our ability to sense its passing should give us the confidence to jettison manufactured clocks in favor of our body clock. The orbweavers express an ease with time travel that we can aspire to. The sociobiologist who encouraged Martha McClintock to study pheromones, E. O. Wilson, wrote, "*Homo sapiens*, the first truly free species, is about to decommission natural selection, the force that made us. . . . Soon we must look deep within ourselves and decide what we wish to become." Perhaps we should look to our eight-legged companions for guidance.

Thomas Jones said, "If we can understand how these spiders phase-shift every morning without harmful effects, imagine what this could teach us. Perhaps these spiders can show us how to avoid the dangers of time desynchrony seen in shift workers. Perhaps they can help us find drugs to beat jet lag or cure time blindness." Darrell Moore also sees promise: "Given that most of our knowledge about circadian rhythms comes from three or four model species and our studies alone have shown a huge variation, nature has so much more to show us." However, Foster implores us to think carefully about where this knowledge might lead us. "It is not far-fetched to imagine that in the next few years, we will learn how to manipulate circadian rhythms and so disconnect ourselves from the natural world." We need to decide now about what we want for our future. "Will it be the time hell of 'Dyschronia'?" he asked: a world, perhaps, of 24/7 soldiers, spacemen, and shift workers, sleeping only a few hours a night. Or will it be "a time paradise or 'Uchronia' for a time-stressed populace?" Only tomorrow will tell.

11

The Bar-Tailed Godwit and
Our Sense of Direction

ACCORDING TO LEGEND, THE MAORI WERE GUIDED TO their island home, Aotearoa, "land of the long white cloud," by the sun, the stars, and a bird they call the kuaka. For years, their Polynesian ancestors had watched this elegant, stilt-legged wader—known elsewhere as the bar-tailed godwit, or *Limosa lapponica*—flying in flocks from the north and heading south. They imbued the fleeting visitor with mystery, creating proverbs in which it symbolized the unobtainable: *Kua kite te kohanga kuaka? Ko wai ka kite I te hua o te kuaka?*—Who has seen the kuaka's nest or ever held its eggs? Those who dreamed of the unknown lands to which it flew would set sail in its slipstream. They became the first people to set foot on what Captain James Cook would later call New Zealand. The Maori could not have known that the time they shared with the bar-tailed godwit was just a moment in one of the animal kingdom's most epic journeys.

In August 2007, US Geological Survey biologists Bob Gill and Lee Tibbitts were studying bar-tailed godwits at the other end of the world. "Lee and I were sitting in a remote Alaskan village," Gill told me. Tibbitts added, "I was logging on to my laptop to see if there was any new data on our birds." Earlier that year, in February, sixteen godwits had been caught in New Zealand and implanted with radio transmitters, enabling satellites orbiting the earth to relay their map coordinates to Tibbitts's screen. "As well as location, they report battery strength," she said. "They are on for eight hours, when we hope for two to three satellite readings, then off for twenty-four to save the battery." They had used the smallest transmitters available at the time. "But even these were 26 grams [just under an ounce], a fair amount of extra cargo for a godwit," said Gill, considering the birds themselves are only ten times that weight at the start of the season. Over the spring, Gill and Tibbitts had tracked their tagged birds flying thousands of kilometers north from New Zealand, via a long rest in China by the Yellow Sea, to their breeding grounds in the Alaskan tundra. In 2002, similar methods had been used to follow the Far Eastern curlew from Australia to China;

they revealed that one bird had flown some 4,500 kilometers (2,800 miles) over water in the world's longest nonstop migration. Gill and Tibbitts wondered whether the godwits might challenge this record as the birds returned south once more to New Zealand. However, that summer they had watched their sixteen travelers fall off radar, as battery after battery died. All bar one.

Gill and Tibbitts were still getting reports from a lone female called E7. Her battery was low but working. Gill said, "I remember Lee and I were thinking aloud at what a cruel joke this was because, based on past performances, once batteries reach the level of hers, we could expect them to fail within a matter of a day or two." Each day, they waited and watched to see whether it held and whether E7 had set off. Then, at 10:00 p.m. on August 29, a report came in that she had flown and was just south of Alaska, over the Aleutian Islands. "We were so excited that we jumped up and down," Tibbitts recalled. The Pacific is the largest ocean, covering 30 percent of the earth's surface. There would be nowhere to stop until Hawaii. E7 flew through two nights toward the archipelago. In the early hours of the third day, when the next report came in, she had made a turn; she was nearly 650 kilometers (400 miles) west of Hawaii and flying past without landing. With 4,800 kilometers (3,000 miles) beneath her wings, she had knocked the Far Eastern curlew off its perch, and she was not even halfway there. By the sixth day, she topped 8,000 kilometers (5,000 miles) and carried on past the Polynesian island of Fiji. "Fiji was a milestone," Gill said. "Again, it meant open ocean until New Zealand, so now she had no choice but to carry on." All hopes now fixed on the bird and her battery.

On the afternoon of the eighth day, two reports came in quick succession. The first showed E7 only 150 kilometers (100 miles) off the most northern tip of New Zealand. Two hours later, the second showed her flying southwest, heading out over ocean. "We were shocked. Where was she going?" recalled Gill. They weren't the only ones waiting for the godwit. "There were crowds on the beach from which she had originally flown, back in March. There were cameras. The world was watching." The vigil continued into the night. Then she was back on air. Gill continued: "It strained credulity, but the satellites had located her by the Piako River." The godwit had landed. "Seven miles [11 kilometers] from where she had first been tagged," Tibbitts elaborated. Since then, E7 had covered 29,500 kilometers (18,300 miles). Her final 11,680-kilometer (7,300-mile) leg, almost a quarter of the way around the world, had been done in eight days without stopping. Three years later, an Arctic tern would be tracked across 71,000 kilometers (44,000 miles) in one year, to claim the longest bird migration. In 2010, a bar-headed goose would soar 10,175 meters (33,400 feet) above sea level while crossing the Himalaya, to claim the highest bird migration. The bar-tailed godwit still holds the record for the longest nonstop hop, by some distance.

"To fly all the way across the Pacific, without rest for eight days, was so much more than we thought possible. No food, no water, no sleep as we know it," said Gill. "I try to be objective as a scientist, but this is a head-scratching, jaw-dropping feat." Research has since shown that the godwit uses various tactics. Ahead of takeoff, its body undergoes remarkable change. Not only do organs such as the gizzard and gut shrink to make space for expansion of flight muscles and lungs, but also vast fat reserves are laid down. "The birds double in mass in just two months," said Gill. "They look like flying balls." "And squidgy to touch," added Tibbitts. The team also discovered that bar-tailed godwits take advantage of prevailing winds. From Alaska, they ride the tailwinds of storms blowing south. "Although they still have to flap their wings continually, the winds boost their speeds to 50 miles [80 kilometers] per hour," he explained. As they head farther south, they hitch onto different wind systems. Gill added, "It was only when we saw the birds moving through wind fields, choosing the best ones for their location, that we realized they must know exactly where they are." Which raises a profound question: How does this ultramarathon champion find its way across the flat and featureless ocean through day and night?

*

In 1873, the journal *Nature*, which had been running for only four years, invited contributions from its readers on the possibility that humans and other animals possess an innate sense of direction—a sense that, like those of balance and time, had eluded scientific inquiry because it works without conscious sensation. Various scientists responded; among them was Charles Darwin. Darwin's paper cited as evidence "the English translation of the Expedition to North Siberia by Von Wrangell [*sic*]," which recounts how the explorer was struck by his Cossack driver's ability to hold a true course through the twisting, hummocky ice fields without need of a compass. A century before this, Captain Cook had seen in the Polynesians a similar knack for navigation. Inviting Tupaia, the dispossessed chieftain of Raiatea, on board HMS *Endeavour*, Cook and his crew were astonished at his ability to point homeward without any navigational tools. With Tupaia's unerring guidance, the ship wended its way, between atolls and archipelagos, toward New Zealand. Such tales, combined with Darwin's words, crystallized the belief that animals, including humans, though perhaps some more than others, share an innate flair for orientation. The notion of an unknown and secret sense persisted throughout the latter part of the nineteenth century and into the early twentieth. Lacking evidence, the idea fell into disrepute until the 1950s, when this changed—for E7's kind, at least.

Ornithologists had long noticed that during the migratory season, caged birds will flutter their wings and become agitated. This irrepressible urge to move became known as Zugunruhe, from the German *Zug* (migration) and *Unruhe* (restlessness). Even before technology offered such wonders as satellite transmitters, one scientist realized there might be a way to study migration. Cornell University's Steve Emlen hatched a simple yet ingenious plan. He constructed a funnel-like birdcage, lined it with blotting paper, and turned its base into an inkpad. When subjects under the spell of Zugunruhe were placed in this contraption, inevitably they would leave a trail of indelible footprints, thereby indicating the compass direction in which they hoped to head. Researchers worldwide were soon using Emlen funnels to explore how birds sensed direction. Early results implied that like the Polynesian mariners, they steer a course by the heavens. European starlings were seen looking to the sun. Emlen headed to a planetarium, with his funnel and a flock of indigo buntings, and found these birds look to the stars. As evidence grew, scientists began to talk of daytime migrating birds having an internal sun compass, nocturnal ones having a star compass, and those that travel through day and night, like E7, both. Neither compass works like a man-made version, as they rely on the sense of vision. However, a student at Frankfurt University was seeing less easily explained behavior in the common or garden robin.

Under cover of darkness, some European robins migrate south in winter; those in northern Germany might fly southwest toward England or onward to warmer climes. Sure enough, Wolfgang Wiltschko and his supervisor, Friedrich "Fritz" Merkel (not the Friedrich Merkel of the touch chapter), found that under a starlit night sky, captive robins hopped in a southwesterly direction. However, they continued to hold this steady course even when thick clouds plunged them into darkness. The birds seemed capable of navigating without recourse to the sun or stars. This observation set in motion a series of experiments that would ultimately prove the existence of a true compass sense in birds.

*

Deep beneath our feet—a plunge of 3,000 kilometers (1,800 miles) but a mere two-day level stretch for a godwit—is a swirling mass of molten metal. Essentially it transforms the planet into a giant magnet with magnetic poles. These poles inexorably draw the needle of a man-made compass, itself a lightweight magnet. Although magnetic north is the name of the pole to which the compass needle points, technically, it is magnetic south pole because opposites attract. To add to the confusion, magnetic north is not what we have come to know as the North Pole; tugged by surges in the liquid iron core, its location continually drifts and is currently somewhere in the Canadian Arctic, hundreds of kilometers away from the northern end of the axis on which our planet spins. The magnetic field that arises between these poles is faint; whereas a small refrigerator magnet exerts 10 million nanotesla, the earth's magnetic force from equator to pole reaches no more than 65,000 nanotesla. By convention, it is visualized as lines. These emerge vertically from one pole, flatten as they loop over the earth to run parallel at the equator, then regain steepness and descend vertically on the other pole, as if enclosing the earth in a birdcage. Of course, in practice the magnetic forces are beyond our sense of sight, as well as smell, touch, taste, and hearing. Wiltschko and Merkel suspected they were not beyond the sensory world of a robin.

The scientists returned to their experiment. This time, they encircled the bird funnels with Helmholtz coils, which, when supplied with electricity, create a local and adjustable magnetic field. When they switched the field's orientation, the birds responded dutifully, switching the direction in which they hopped. Somehow the robins were able to read and respond to the weak magnetic field. Eight decades after *Nature*'s original appeal to scientists, here was evidence of a sense of direction. More experiments and more revelations followed. The names of Wolfgang and Roswitha (his wife) Wiltschko became synonymous with the novel perception of magnetoreception. Emlen's indigo buntings joined the flock, as did pigeons, blackcaps, and warblers. Today, scientists suspect a hidden compass guides all birds. Recently they discovered it has a level of sensitivity beyond their wildest imaginings.

"This experiment was completely unplanned," said Henrik Mouritsen. He runs a unique facility at the University of Oldenburg in north Germany, dedicated to the study of night-migratory songbirds. "There is nowhere like it in the world," he told me. It came about through necessity. When Mouritsen first arrived from Denmark via Canada, he set out to replicate the famous Wiltschko experiment. "These funnel experiments showing birds orienting to the magnetic field had been done in fifty different labs around the world." They had been repeated so often that the test had become standard. So as his first Oldenburg winter approached, Mouritsen was ready with a round of robins. The research did not go to plan. "For five migratory seasons in a row, I had PhD students putting robins into funnels every single night without success." Every morning for five years, during the months of Zugun-

SENTIENT

ruhe, the students would check the Emlen funnels in the windowless
wooden huts but found no hint of directionality in the trails left by
the birds. "They were completely random," he recalled. "It was highly
frustrating, for me and my students." They tried adjusting all manner of
parameters. "The food, the size of the funnels, their shape, the light intensity, the daylight cycle," he listed. "We even gave the birds vitamins.
We tried everything we could think of, and still no effect." Mouritsen
was at a loss. Then a student proposed using a contraption that is more
often chosen when recording nerves firing: a Faraday cage. Mouritsen
was hard-pressed to see the relevance this device might have to magnetoreception. "Normally, I'd have said this suggestion was nonsense,
but I was desperate." He ended up transforming all the Oldenburg bird
huts into Faraday cages: walls and ceilings were lined with thin sheets of
aluminum and grounded with a lightning conductor. The effect on the
birds' behavior was immediate. "It was like a miracle," he told me. "The
birds oriented." When Mouritsen turned on the Helmholtz coils and
flipped the local magnetic field, the birds flipped direction, as they had
done for Wiltschko years before. Yet when the next migratory season
arrived, the experiments failed again.

"I couldn't believe it. I couldn't waste another season," recalled
Mouritsen. A new student was conducting the testing. After a few
weeks without results, Mouritsen decided to follow her as she set up.
"It looked like she was doing everything right," he said, "that is, until I
went around the hut and noticed that she had forgotten to connect the
grounding. Basically, she had omitted to turn one screw." This gave him
an idea. He saw an opportunity to conduct a double-blind study, the
gold standard for eliminating bias in results, in which the experimenter,
as well as the subject, is oblivious to the control procedure. So without
telling her about the screw, he asked the student to proceed as usual,
now testing birds in two huts. "What she didn't know was that every
two days, I connected the grounding on one hut but not the other. Then
I went in again and switched them," he said. "When I asked her how the
birds were doing, she tried to be positive, but she found their behavior
confusing." The results were unclear to her, but they were crystal clear

to Mouritsen: the birds could be made to alternate between orientation and disorientation at the turn of a screw. "It was extraordinary," he said. "I was happy to have located the problem but mystified. Theoretically, this should not have been happening." It would take seven more years of research until Mouritsen felt he could prove, beyond doubt, what had been hijacking the birds' compass sense. When his conclusions were published in *Nature*, he warned, "My first reaction was 'It can't be.' The first reaction of most people to this paper will be 'It can't be,'" but it is." And the culprit is not only ubiquitous but also man-made.

Birds orient when the Faraday cage is grounded and activated because it strains electromagnetic fields selectively. It allows access to the forces created by the earth's natural magnetic field, while blocking those from unnatural electrical technology. "Our results show that the bird compass is disrupted by the electromagnetic fields that are generated by a wide band of radio frequencies," said Mouritsen, "though it remains unclear how this happens." The team proved that the offending frequency band is from 2 kilohertz to 5 megahertz, implicating all sorts of everyday devices, from computers to televisions, refrigerators, and kettles, as well as medium-wave AM radio signals. He added, "These signals are common throughout cities, towns, villages. In fact, you can listen to the radio in your car almost anywhere, even in the most remote places." The California Institute of Technology professor Joe Kirschvink wrote a response to the study under the title "Radio Waves Zap the Biomagnetic Compass," questioning whether we have been unknowingly addling bird brains since the first AM radio transmission over a century ago. "But there was a contradiction," said Mouritsen. Every year E7 and her kin venture out over the Pacific and find New Zealand; flocks of other species crisscross the globe with accuracy. "Migrating birds were still reaching their destinations."

"I thought it would be impossible to find a place without these signals, but it turned out that I was wrong. I had only to drive 1 kilometer [0.6 mile] outside the city." By running the test in an old wooden stable just outside Oldenburg, Mouritsen and his team saw that the birds could orient again without the need for a Faraday cage. This implied the electromagnetic noise from radios and all other electrical gadgets was disruptive only at source, so they returned to the Oldenburg huts and measured the local levels. "The earth's magnetic field on campus is about 49,000 nanotesla, but we started seeing birds being disrupted by electromagnetic noise from any single radio frequency of only 50 nanotesla." They were reacting to a man-made force barely one-thousandth of the earth's magnetic field, itself quite weak. "Eventually we found that birds stopped orienting with only 1 nanotesla of any single frequency." Mouritsen and his team had proved that the bird compass can respond to forces that are one fifty-thousandth of the natural magnetic field. Such sensitivity shocked the scientific community. Kirschvink told me, "The levels are substantially below anything previously thought to be biophysically plausible." Mouritsen said, "Remember we only discovered this by chance, but the evidence is clear. The bird's magnetic compass is a million times more sensitive than any other sensory system known." Birds are alert to infinitesimal shifts in the earth's magnetic field.

As E7 set off on her long-haul flight over the Pacific Ocean, she would have been guided through night and day, regardless of the weather, by the help of an extraordinarily keen magnetic compass. Perhaps she detected the angle of the force lines reducing as she approached the equator; perhaps she sensed their waning intensity as she pressed on. Either would have informed her of her north–south position, and some evidence suggests that birds use this latitude to calculate longitude, allowing them to pinpoint their coordinates. In addition to a compass, she would also need some form of internal map; only through knowing where she was could she decide in which direction to head. "How the magnetic sense works is perhaps the most significant unresolved question in sensory biology," said Mouritsen. Crucially, it is the final sense whose sensor has yet to be identified. Over the past decade, molecular biologists, electrophysiologists, neuroanatomists, geophysicists, and quantum physicists have joined the hunt, but the magnetoreceptor remains elusive. The challenge facing scientists is that because the earth's magnetic field penetrates living tissue, the sensory organ need not be visible on the surface of the bird. The sensor might not necessarily be an organ but a scattering of magnetoreceptors. There are two rival theories, each with compelling evidence in its favor.

＊

Scientists such as Henrik Mouritsen look to the most understood organ, the eye. They believe it hides yet another secret sense in addition to that for time. Invoking the veiled world of quantum mechanics, the study of how the fundamental particles of the universe work, they argue that the magnetic field triggers quantum reactions in a group of light-sensitive proteins called cryptochromes, which are found in the retinal rods and cones of a bird's eye. "Cryptochromes are the only candidate we've found in the last twenty years that fit the bill," said Mouritsen. "Although we haven't looked at godwits specifically, it is reasonably safe to assume they have them. Cryptochromes have been found in the eyes of all the birds studied so far." When a photon hits rhodopsin, its retinal molecule snaps into a different shape, setting in motion an act of sight; the quantum compass theory proposes that when a photon strikes a cryptochrome, it generates free radicals with mismatched pairs of electrons. "As electrons are charged and moving, they are in essence microscopic magnets," explained Mouritsen. Consequently, their magnetic fields could interact with the earth's magnetic field. "These radicals only last for one millionth of a second. We propose these tiny and brief interactions have profound and lasting effects in the birds' eyes. We think they enable the bird to actually see the earth's magnetic field." The quantum compass theory endows birds with a new channel of sight: a real-world equivalent to superhero magnetovision. While regular cones and rods might paint a panorama that we can recognize, specialized magnetosensitive rods or cones act to overlay it with a magnetic compass halo. Mouritsen explained, "Midway across the Pacific, E7 sees an aerial view of the ocean, but perhaps this is superimposed with some form of shading that illuminates which way is north." This is one theory. The other has nothing to do with fleeting free radicals and elusive electrons. It proposes an earthbound compass and mines a rich seam between biology, physics, and geology.

"I am a biophysicist, a geobiologist, a geophysical neurobiologist," Joe Kirschvink informed me. "Frankly, I don't know what to call myself." He has also earned the title "the real Iron Man" because he is the main advocate for the existence of an animal compass that is made from the iron mineral magnetite. "Magnetite is my favorite material," he explained. "It is naturally magnetic, and although regularly found by geologists in igneous rocks, it is also found in living creatures." Decades ago, certain mud-loving bacteria from the tidal flats of Woods Hole, Massachusetts, were found to be magnetosensitive. When scientists looked through their transparent membranes, they saw tiny curved chains of magnetite crystals. "The individual crystals are too small to see with the naked eye. They are about 50 nanometers long so you'd need a line of 20 million or so to reach a meter [3 feet]," said Kirschvink, "but thanks to them, anyone can command their movement. All it takes is a small bar magnet." One laboratory even manipulated millions of these bacteria to make them line-dance in time to the tune of "Cotton-Eyed Joe." Effectively, magnetite turns these creatures into a shoal of swimming compass needles. Next, scientists found magnetite crystals in the abdomens of honeybees and the beaks of birds. Kirschvink and his collaborators argue that the crystals are part of a biological compass organ, similar to the cones of the eye and the hair cells of the cochlea, which in this case responds to the earth's magnetic field rather than to light and sound. "As the crystals align to magnetic north, their movement could trigger surrounding mechanoreceptive structures to fire off nerve signals to the brain." According to this theory, there are potentially millions of such magnetite crystals acting like microscopic compass needles, thereby enabling the brains of bees and birds to process their bearings.

Whether the bird's compass is quantum or mineral in essence is the subject of heated debate. "The compass has to be quantum mechanical in nature because otherwise it is impossible to get anywhere near the sensitivity that I saw with the radio frequencies," Mouritsen told me. "Twenty years ago, Joe said we were seven orders of magnitude away from proving the quantum compass. I think we are now two, and I'm confident the next decade will close the gap." Kirschvink demurred: "If anything, I'd say the quantum theory is even further from being proven. Its advocates have published many papers on molecular findings, but they do not explain what we see." He challenges Mouritsen to explain how, for example, a quantum compass that relies on light can account for the many animals that migrate in the dark: birds flying through the night, whales migrating in the deep ocean, salmon traveling under the polar icecaps. Adding: "The quantum theory unnecessarily overcomplicates what the magnetite theory explains quite simply." In answer, Mouritsen argues that the burden of proof falls equally on Kirschvink. How, for example, can he be sure that the magnetite discovered is sensory and not part of some other biological system? Perhaps it is the body's way of storing essential minerals through iron homeostasis.

The argument descends into more detail, quoting and questioning other lines of evidence. That said, Mouritsen and some others wonder if the two theories need not compete. "There is evidence that many birds have both a magnetic compass and a magnetic map, so the two senses could work together. If the cryptochromes allow the bird to see the magnetic field and act as the compass, the magnetite crystals could provide the map." This might mean that birds possess not one but two distinct senses of direction. Whatever the sensor, or sensors, the facts are fast becoming stranger than fiction—leading one scientist to exclaim, "You couldn't make this stuff up"—and it seems the sense is not solely the preserve of the bacteria, the birds, and the bees.

*

The more scientists look, the more they find proof of a compass sense elsewhere in the animal kingdom. Magnetic sensitivity has been observed in featherless long-distance migrants: from loggerhead and leatherback turtles to Pacific salmon and glass eels. It has been seen in feather-light Australian bogong moths and monarch butterflies, whose annual 5,000-kilometer (3,000-mile) round trip between Canada and Mexico is one of the wonders of the natural world. It might also be present in the less well-traveled eastern red-spotted newt and western Atlantic spiny lobster, as well as the near-sedentary tidal mud snail

Nassarius. Some studies suggest that mammals align themselves to the earth's field as part of their everyday routine. Roe deer might orient their heads northward when grazing, blind mole rats are prone to siting nests in the southernmost wing of their sprawling underground labyrinths, and "man's best friend" has been seen tending to a north–south axis when evacuating its bowels. "Some of the scientific inquiries are more solid than others," Mouritsen told me, "but it is fair to say we now think this sense is not present in just long-distance migrants." Kirschvink suspects the sense starts in deep time. Magnetite chains have been identified in fossil bacteria that are 2 billion years old. "The sense is even more ancient," he told me. "Subsequent research shows that it likely predates a major split in the bacterial domain, meaning it was in the ancestors to nearly all living organisms. Magnetoreception is a primal sensory sense and will be found throughout the animal kingdom." No wonder one scientist set his sights on finding it in a larger and somewhat more sophisticated creature.

Over a period of three years beginning in 1976, the University of Manchester staged some extraordinary experiments on the nonmigratory, desk-bound British student. Robin Baker, who at the time was a lecturer at the School of Biological Sciences, persuaded pupils in his behavioral ecology course to turn guinea pig and subjected them to a method long used by zoologists: the tried and tested displacement-release experiment. He invited volunteers onto the roof of the Zoology Department to get their bearings. Then he herded them blindfolded into a Sherpa van and drove them on a circuitous route deep into the Pennines. Upon stopping, he asked them to point home, as centuries before Captain Cook had of Tupaia. "The vast majority considered themselves to be lost," he later wrote, "and were always genuinely surprised when their group estimate turned out to be accurate." The results were so compelling that he proposed his students must be drawing on an innate sense of direction. Next, he needed to ascertain whether it involved magnetism.

Baker commissioned specially designed helmets incorporating the Helmholtz coils that Wiltschko had used, and Mouritsen would use, on robins. "These helmets supported two lateral coils, each of 200 turns of 40 swg [standard wire gauge] copper, covering the braincase," he explained. Switched on, they generated a magnetic field three and a half times greater than that found in nature. Assembling another group of students, Baker put half in the Helmholtz helmets and the rest in others that looked and felt identical but were magnetically inactive. All were blindfolded, displaced, released, and once again asked where home lay. "The pattern that emerged was as dramatic as it was unexpected," Baker recalled: those wearing the Helmholtz helmets were less likely to point in the direction of home. He surmised that their ability to judge direction had been disrupted, which "in turn implied that people have a magnetic sense of direction." After 140 volunteers and 940 individual releases, Baker felt persuaded to posit that all humans, whether Tupaia or a less-practiced student, share some of the godwit's gift for navigation. He drew comparisons with the senses of balance and time, which work silently and secretly within us. "When we compare the senses of time and magnetic direction, the parallels are striking. . . . Both are relatively crude, yet adequate. Perhaps both can be trained to greater accuracy." He concluded, "At first the possibility that humans have a magnetic sense seems incredible," but consider its prevalence in the natural world: "In this wider perspective, it would be incredible if man did not have a magnetic sense." He set about publishing. "The first two papers on this work were sent to *Nature*, more than one hundred years after the journal's initial invitation," he recalled. "Both were rejected." Eventually one of them was picked up by the rival publication *Science*. The research community was quick to react, and some labs set out to replicate the results.

"When I first saw Baker's work, my first reaction was, 'Oh God, this makes sense,'" Joe Kirschvink told me. "I was at Princeton at the time, and we asked him over to run the same experiments." Baker accepted the invitation. The magician and paranormal skeptic James Randi even offered himself up to blindfolds and disorientation. "But no one, not even the great man himself, could repeat Baker's results," Kirschvink said. Eight experiments later, the Americans published under the title

"Human Homing: An Elusive Phenomenon," claiming, "These results provide no support for the hypothesis that humans can determine direction of displacement or sense the earth's magnetic field." Then he added, acerbically for an academic paper, that perhaps Brits "are dramatically better than Americans in using whatever cues may be involved." Five years later, and 60 kilometers (40 miles) from Baker as the crow flies, a team from Sheffield University reran the experiments. They summed up their efforts with the words, "Human homing: still no evidence," asking, again somewhat drily, "Perhaps it depends on which side of the Pennine Hills the experiments are conducted?" Baker persevered, trying different experimental approaches and publishing books such as *Human Navigation and Magnetoreception*. Nonetheless, the ensuing decade became mired in what he called "a series of transglobal, trans-Pennine and particularly transatlantic animosities" that ultimately "descended into personal defamation." In the end, Baker abandoned his search for the human compass, saying, "I have no doubt that the conclusion will still be resisted for one reason and one reason alone: the sense is an unconscious one and in the minds of many people, therefore, cannot exist." For four decades, the field lay fallow, until chance intervened.

*

A few years ago, Joe Kirschvink was attending a Caltech faculty party. "They are great melting pots, so I always try to go," he told me. "That time I was introduced to Shin." Shinsuke Shimojo is the professor in charge of the Psychophysics Lab. Kirschvink, bitten by the magneto-bug, had been continuing the hunt. "I started telling Shin about the frustration I'd had trying to repeat some of Baker's experiments. The thing was that occasionally I saw really significant data. For example, one student would get the direction right fifty out of fifty-five times, but would then come back the next day, unable to repeat it. There was just so much variability to it all." Shimojo's expertise offered a different viewpoint. "Shin asked if I'd thought to look at the brain. 'Joe,' he said to me, 'behavior is a conscious perception. If there really is a magnetic sense, even if this sense is unconscious, the brain has to respond.'" This insight changed everything. Kirschvink did not need to locate the magnetoreceptors; he joked, "The magnetic receptors could be in your left toe and the sensory information all ends up at the brain." Baker's human behavior tests had been beset by all sorts of variables beyond his control; his results had depended on a subject's motivation, focus, and attention, even on that person's previous experience and memories. Kirschvink realized that by targeting the brain, he could eradicate this behavioral "noise." He said, "Shin's suggestion of this direct physiological approach was brilliant." Now the challenge became one of implementation.

Two floors beneath the Caltech Division of Geological and Planetary Sciences is Kirschvink's Human Magnetic Reception Laboratory. Having followed the trials and tribulations of Henrik Mouritsen, he believed there was much to learn from the Oldenburg robins. "The Faraday cage is key," he told me. "Now, I wonder whether electromagnetism interfered with our first attempt to repeat Baker's work all those years ago." Kirschvink decided to design and build a human-sized Faraday cage. "By May 2015, I had three undergraduate volunteers assembling a cage about 2 meters [6 feet] cubed in the Mag Lab." The human test chamber, like that of the robin, was lined with grounded aluminum to shield it from external man-made electromagnetic noise, as well as from light and sound. Acoustic panels further deadened sound from within. It was surrounded with Merritt coils, much like Wiltschko's Helmholtz coils, to generate a uniform magnetic field, but this time over a larger area. "We used these to create a weak magnetic field around 35,000 nanotesla, which is what we experience every day in our laboratory, outside the test chamber," Kirschvink said. "We also built it so we could vary the direction, as well as the intensity, of the magnetic field." They arranged three sets of four coils perpendicular to one another to create forces in three dimensions: not simply in the north–south and east–west axes but also up and down. "This meant we could rotate the magnetic fields about the subject so they experience changes in magnetic stimuli as if they were turning their head and moving, when in reality they are sitting still." They wore a skullcap studded with sixty-four electrodes so any activity in their brain would be monitored through electroencephalography (EEG). Kirschvink realized that in order to attribute any brain responses to shifts in magnetism, they had to eliminate all other sensory experiences. So in addition to seating the subjects in a reclined armchair, they also plunged them in total darkness and near silence. The absence of light would have the added benefit of eliminating the quantum compass theory from the results. Kirschvink recalled his own first experience in the hot seat: "It was so dark, quiet and boring, the chair so comfortable, that the difficulty was not falling asleep. I felt nothing at all." Yet his brain was being monitored constantly as the magnetic fields shifted restlessly around it.

"From day one, we had interesting responses," Kirschvink said. "Our first volunteer was in the cage. We had run a few controls with the fixed magnetic field and were looking at his EEG reading on the monitor as the fields changed." His test hour in the cage was nearly up. Already, he had been subject to several seven-minute trials. In some, the experimenters had kept the field fixed; in others, they had changed the direction of the magnetic field approximately one hundred times at irregular intervals. Any EEG differences between the two sets of data would indicate that his brain was processing geomagnetic stimuli. Then one of Shimojo's graduate students, Connie Wang, started to notice a drop in his brain's regular pattern of alpha waves. These followed certain field rotations. The researchers wondered whether they were witnessing a phenomenon known technically as alpha event-related desynchronization. Kirschvink explained, "It is how the brain responds to sensory stimuli like a flash of light, a stroke on your arm, or an auditory tone." It is so well documented that Shimojo calls it "the signature for sensory detection." When we are awake and relaxed, the spontaneous activity of the millions of neurons in our brain creates alpha waves on the EEG that oscillate at a rate of ten per second. "However, this soothing alpha hum is lost when there is a sensory input," Kirschvink told me. "If this were alpha desynchrony, it indicated that the volunteer's brain was picking up and processing a sensory stimulus." Given the cage had been designed to exclude all sensory signals bar one, it could only mean that the volunteer was responding to magnetic shifts.

The Mag Lab team pressed on, testing thirty-four people. Many months later, they were in a position to amalgamate and analyze the data. The early indications had persisted. "The results revealed that half a second after a magnetic field rotation, most people had a significant alpha wave collapse. Some dropped by up to 60 percent," Kirschvink said. "We concluded this is evidence of the human brain processing earth-strength magnetic fields. It means that humans have a compass sense, just like the birds and the bees." With a dig at Mouritsen, he added, "Because the experiments took place in the pitch black, without light, the results show that the compass cannot be quantum." When the volunteers were asked about the experience, not one reported having felt a thing. "This sense works in us unconsciously, without our awareness," explained Kirschvink. "The interesting question is not whether we sense magnetic fields consciously, but whether we use the information." One of the researchers said, "I was just blown away by the results. I didn't ever think we'd find anything this clear-cut, quantifiable, and reproducible." When the paper was published and posted on the Caltech website, Kirschvink received a surprise call: "The tech team were on the phone in a panic. They had been inundated with hundreds of requests to download the study data and it was clogging the system. 'Joe,' they said, 'change the link, the system can't cope.' I think it is the first time a scientific paper has crashed Caltech's website."

The reaction to the paper has been mixed. Skeptics question whether the brain wave dip is necessarily evidence of a hidden sense. One dismissed Kirschvink's conclusions, saying, "If I were to stick my head in a microwave and switch it on, I would see effects on my brain waves. That doesn't mean we have a microwave sense"; from another: "If we could sense the earth's magnetic field, we would probably know about it by now." Not long after, Robin Baker resurfaced with a thirtieth-anniversary publication of his book. "I am delighted the field appears to be stirring again," he admitted. "The re-emergence of Joe Kirschvink, an old adversary, to cheekily claim (or maybe, in his defence, to have others cheekily claim on his behalf) that he is on the verge of discovering the existence of human magnetoreception triggered a wry smile." Mouritsen was more circumspect: "Joe's experiments are provocative. The procedure is robust. The data is indirect, but interesting. But I do not think the results are fully consistent with Joe's interpretations that it must be a magnetite-based mechanism. More experiments are needed before we can understand what they exactly mean." A team in Tokyo has taken up the challenge. Meanwhile, Kirschvink is navigating new terrain. "My feeling is that this discovery is just the tip of the iceberg. Yes; we need to replicate the study. Yes; we need to find the sensor. But what if some people are aware of this sense?" NASA has funded Kirschvink to take the Faraday cage to Australia. He plans to study a remote tribe whose people seem susceptible to the spell of Zugunruhe and in possession of a remarkable talent.

*

The Guugu Yimithirr are indigenous people from the outback of northern Queensland. Their first claim to fame came in 1770, when they taught Captain Cook the word *kangaroo*. More recent research frames the Guugu Yimithirr as the human version of godwits; not only do they heed the call of the walkabout but they do not get lost. In displacement-release experiments, the Guugu Yimithirr have been driven and walked long distances, through twists and turns, taken to dense forests, even caves. Yet always, without hesitation, they can point accurately home or in the direction of north, east, south, or west. The linguist Guy Deutscher described them as having "an almost superhuman sense of orientation." This faculty has been seen in other indigenous Australian tribes and long attributed to language. "We are going to investigate a group of people called the Girundji," said Joe Kirschvink. Like the Guugu Yimithirr, they lack words for left and right. "Instead, they describe their world with the cardinal points. They say things like, 'I stubbed my north toe' or 'I have a fly on my south shoulder.'" Similarly, according to the cognitive scientist Lera Boroditsky, the way you greet someone in another indigenous dialect, Kuuk Thaayorre, is to ask which way someone is going. "The answer might be, 'North-north-east in the far distance. How about you?'" she said. "Imagine as you're walking around your day, every person that you greet, you have to report your heading direction; that would actually get you oriented pretty fast because you literally couldn't get past hello if you didn't know which way you were going." Language and conversation clearly ensure a constant awareness of their orientation, but this does not explain how it is first perceived.

Deutscher has suggested these tribes have a compass of the mind "that operates all the time, day and night, without lunch breaks or weekends." When asked to explain their directional skills, they cite clues in the landscape; some look to the sun like starlings, some to the stars like indigo buntings, some to the direction these winged migrants fly to or from. Yet further elucidation evades them. "They cannot explain how they know the cardinal directions, just as you cannot explain how you know where in front of you is and where left and right are," Deutscher observed. "They simply feel where north, south, west, and east are." Kirschvink is convinced that their cues are not simply visual and Deutscher's compass is more than metaphor. He suspects these people sense the earth's magnetic field and do so more acutely than the Californian Mag Lab volunteers. "They might even be conscious of the earth's magnetism," he told me. "I wonder whether these Australian tribes can show us how to unlock our own potential. Perhaps they can even transform an unconscious possibility into a conscious reality."

Of all the senses explored in this book, the sense of direction is the only one that is yet to be proven in humans. Like the Maori forefathers, science is striking out into the unknown. Studies have revealed how the ancient mariners would have been guided by the sun, the stars, and also the earth's magnetic field. They would have adjusted their path, through night and day, under clear skies and clouded, to this compass sense. The question is whether they were simply following the ancestors of the intrepid kuaka E7 or listening to something deep within each of them. The answer is surely over the horizon.

12

The Common Octopus and Our Sense of Body

I N April 2016 news of a daring prison break hit the headlines. The London papers hailed "The Great Escape," those in New Zealand declared "Inky's Done a Runner," and the *New York Times* reported that "the keepers noticed the escape when they came to work and discovered that Inky was not in his tank." Inky was none other than a New Zealand common octopus who had been the star attraction at the country's national aquarium. "I was really surprised," recalled the aquarium manager, "but then, this is Inky. He has always been a bit of a surprise." Staff learned that the aquarium lid had been left ajar during maintenance work. They found evidence of a slippery trail. After lockup, it seemed that Inky had seen the chink and made a break for freedom. He had suckered up the inside and down the outside of his glass enclosure, slithered across the wooden floorboards, and made his getaway down a drain whose 50-meter (160-foot) pipe led eventually to the big blue beyond of the Pacific Ocean. "I don't think he was unhappy with us, or lonely, as octopuses are solitary creatures. But he is such a curious boy. He would want to know what is happening on the outside," said the manager, adding wistfully, "He didn't even leave us a message."

Octopuses are renowned escape artists. Unlike Inky, most are thwarted, found hiding in all manner of places from the tops of bookshelves to the insides of teapots. Like Inky, all display an uncanny knack for bolting when their keepers' eyes are turned. Scientific research is now starting to reveal the remarkable extent of an octopus's wiles. They learn to navigate mazes, dismantle Lego sets, and even recognize faces; one notorious captive would unleash its water jet with startling accuracy only at one particular keeper. Octopuses can uncork bottles, open screw tops from outside or inside jars, and, impressively, tackle childproof caps on medicine bottles that can bamboozle humans with university degrees. Yet perhaps the single most surprising aspect to Inky's great escape was how he squeezed his eight arms and a rugby-ball-sized body through a drainpipe. The octopus is nature's Houdini.

The Hebrew University of Jerusalem is an improbable home to

forty common octopuses. Living in rows of saltwater tanks, they outnumber the laboratory scientists two to one. "In the early days we had a lot of escapees and they often ended up in other tanks, or drying out on the floor," said Benny Hochner, professor of neurobiology. "We realized they could lift and maneuver the tanks' heavy lids, so now we tie them tightly shut." Hochner's Octopus Lab is part of an international effort to build the world's first soft-bodied robot. His research has attracted the US Navy and the US Defense Advanced Research Projects Agency, both of which hope to use such robots for surveillance and for search and rescue. More recently, the medical profession has expressed an interest in the idea of introducing them, in miniature, into operating rooms. Where some people turn to Bach, Tolstoy, or Rothko for inspiration, Hochner looks to *Octopus vulgaris*. Whether navigating an intricate maze of pipework on a spy mission, sifting through the rubble of an earthquake for survivors, or conducting surgical procedures deep within the veins and tissues of human bodies, Hochner believes that soft-bodied octobots will reach places inaccessible to their stiff, steely edged counterparts, and do so without weakening the surrounding structures.

"The octopus is fascinating," Hochner told me, "the most flexible animal known, with the richest variety of behaviors and senses." Their body has been described as "protean, all possibility," with "less of a fixed shape than any other." Because it is boneless, an octopus can extrude its entire mass through cracks, constrained only by the size of its parrot-like beak. Its arms move much like our tongue or an elephant's trunk, using opposing muscle sets that work against one another. To pass through a hole or make a getaway through a drainpipe, first the muscles that span the arm's width contract, both narrowing and lengthening the limb. Then, on the other side, lengthwise muscles contract, shortening and thickening it again. "An octopus arm can bend at any point and in any direction. It can elongate, shorten, and twist," said Hochner. "Human arms are limited by the bones of our skeleton, but theirs have virtually unlimited degrees of freedom." Such infinite potential enables them to flex to myriad purposes: whether standing, swimming, or crawling; grasping, fetching, or probing; building, digging, even running its suckered arms over its body to groom itself of dirt and skin parasites. Presumably it also demands regimented and rigorous bodily command. Mastering our four skeletally constrained limbs is complication enough: imagine an octopus's seemingly unconstrained *októ*. "For years, scientists have been attracted by the octopus's ability to learn, memorize, and even solve complicated problems," Hochner said, "but I am interested in how they control their highly maneuverable arms and coordinate all eight." Such control and coordination come from a sense that starts deep within the fabric of their being.

In the 1950s, as a young man, the Italian anatomist Pasquale Graziadei spent his summers at the Stazione Zoologica in Naples, where nearby reefs conceal a profusion of suckered tentacles and local trattorias tout a wide variety of *pulpi*. Hooked, Graziadei embarked on an encyclopedic study into the inner structure of the common octopus, an endeavor that would take years but threw light on the extent of its sensory possibilities. We know now that the octopus boasts about half a billion nerves throughout its body, and half of them are sensory receptors. These include mechanoreceptors that give the octopus a sense of touch and chemoreceptors that enable it to "taste" the world through its skin and suckers. Graziadei identified another type of sensory receptor. With his eye pressed to the microscope, contemplating a cross-section of the animal's arm, he saw stellate cells with fine branches extending into and embedded within its musculature. Their fine tracery led back, via various nerves, to the arm's main nerve highway. Graziadei suggested they were sensory cells called "muscle receptors," which fire when stretched as part of the surrounding tissue. Half a century later, Hochner and colleagues showcased this cephalopod muscular sensory system at work for the first time.

"We did these experiments in Naples, in the same place where Graziadei had done his pioneering work," Hochner told me. "He had already mapped out the muscle receptor morphology, so we decided to look at them with electrophysiology to record their activation." The researchers implanted filamentous stainless steel electrodes into the tiny nerve fibers that emerge from the octopus's arm muscles and lead to its main nerve cord. As they moved the animal's limb, sure enough, the electrodes detected repeated and strong excitation. The muscle receptors were responding to being stretched by generating an instant electrical charge that shot along the nerve roots. These sensory cells possess special membranes that become more porous when elongated, thereby transducing physical action into an electric potential. "Our results confirmed Graziadei's theories and showed the stretch receptors firing in real time," Hochner explained. "We found that the most sensitive area, the area that led to the strongest nerve activation, was the outer margin, the first millimeter of the arm musculature." This indicates that the receptors are in the muscle just below the skin. "If you think about it, this is the perfect position to transform movement into electricity. It's the area that deforms most when the arm bends," so it sensitizes the animal to the smallest of movements.

※

Such sensors are not unique to Inky and his extended family. Even before Pasquale Graziadei studied his favorite cephalopod, scientists had found them in creatures that soar, slither, or simply walk. "Stretch-activated muscle receptors are in every animal that moves," said Benny Hochner, "including you and me." Humans possess some 20,000 spindles, buried within and throughout the skeletal muscle of our bodies. They perform the same function as in Inky; at rest, they generate a trickle of nerve impulses; then they too fire rapidly when stretched. Ultimately these muscle receptors give rise to a sense so elusive that it was neither conceived of nor named until the turn of the twentieth century. Many remain oblivious to its existence.

Alongside balance, time, and—the as-yet-unproven—direction, this is another of our secret senses. Not only are its receptors out of sight, but their resulting sensations are so automatic, so familiar that we barely heed them. In the 1890s, the neurophysiologist who first named our sense of pain became curious as to how he could be aware of his limbs even when he lay relaxed and unmoving. Sir Charles Sherrington coined the terms *proprioception*, from the Latin *proprio* (one's own) and *capere* (to grasp), and *proprioceptors* for the stretch sensors within all muscle, whether worm, octopus, or human. Although he gave this core sense of body its name, he was certainly not the first to ponder its existence. After all, three centuries before, Shakespeare's Hamlet had said, "Sense, sure, you have, Else you could not have motion." It is this sense that enables us to feel our flesh as our "property." It acts like an onboard anatomical GPS, locating where we exist in space with such precision that we can guide our finger to the tip of our nose with eyes tight shut, an action that is often used to test whether this sense is in working order. As Oliver Sacks noted, the philosopher Ludwig Wittgenstein equated the resulting unquestionability of our body with the basis of all knowledge. His treatise *On Certainty* begins with the words: "If you do know that *here is one hand*, we'll grant you all the rest." With trepidation, Wittgenstein proceeded to wonder what a lack of such knowledge, such certainty of body, might mean. Dictionaries list terms for blindness and deafness, but do not have a word for being blind or deaf to our body. Perhaps the loss of this sense eludes language because it is unimaginable. In fact, Sherrington declared that such an experience would be "singularly individual and doomed to remain indescribable." Yet this is precisely what befell a nineteen-year-old from Portsmouth, England, one day in May 1971.

"I was at the butcher's where I worked and looked so ill that my boss made me go home," Ian Waterman told me. He seemed to be coming down with a severe bout of gastric flu, but rather than rest, he decided to mow the lawn. "The mower was a fairly slow and ponderous thing but when I got it out of the shed, I couldn't keep up with it." He watched as it careered across the garden without him and came to a halt on the gravel path. Giving up, he went to bed. By the next morning, his condition had worsened. "I tried to get out of bed and fell," he recalled. "Inside I was beginning to panic." His landlady called a doctor, who sent him straight to the hospital. By now, his speech had started to slur. The nurses wondered whether he was drunk. Waterman said, "I had this odd tingling sensation in my hands and feet, but I also realized that

I couldn't feel them. They were numb." He was put into a hospital bed and had the disconcerting feeling that his body was floating above the mattress. As fever took hold, he fell asleep. "I woke up to find a hand across my face. It scared the life out of me. Then, worse still, I realized that this hand was mine." Waterman had woken to a nightmare, the one that had once troubled Wittgenstein and Sherrington: he could not feel his body from the neck down or move it. But the loss was more profound than the numbness of an anesthetic. With his eyes closed, he had no sense of where any part of his body was. He felt as if he had lost himself. Recalling that moment, he told me, "It was as if I had been disembodied." The hospital staff subjected Waterman to reflex tests, blood tests—all manner of medical examinations. They were mystified; they had never seen a case like it. "When the doctors realized there was nothing they could do, they gave up," he said. After five weeks, he was discharged from the hospital, into the care of his mother and father. "I had been left on the scrap heap. That was when terror and fear set in." As he lay on the bed in his childhood home, he brooded over the words of more than one senior neurologist. They had declared that he would never walk again. Thankfully, their prognosis would be proven wrong.

Twelve years would pass. In summer 1983, Waterman knocked on a door in the Wessex Neurological Center and strode into the office of Dr. Jonathan Cole. The English physician, once student, then close colleague of Oliver Sacks, said, "I had never seen, nor imagined, a condition where a person might have a complete loss of touch and position sense. Then, the idea that they would stand, walk, drive, and live independently was astonishing." Waterman seemed to have reclaimed normal life. He had a house, a car, even a job working for the civil service, but appearances were misleading. His every waking moment remained a struggle. Cole would be the first person since that fateful race to the hospital to give Waterman a name for his condition, the first to offer an explanation for how the gastric flu virus had laid waste to his body, and the first to reveal how he had flouted his predicted future.

"From the start, I knew Jonathan was different from all the other doctors I'd encountered. He didn't sit down in his chair on the other side of the desk, but sat next to me, on his desk. He asked me how I coped and what it felt like." Neither man knew then that this would be the start of a friendship that lasts to this day. Cole set about a rigorous study of Waterman's body and became, in his words, "Ian's introduction to neuroscience." Much as Hochner had used electrophysiology to interrogate the octopus's nerves, Cole inserted fine needle electrodes through Waterman's skin and into the muscles of his limbs to test nerve responses. He passed small electric currents through Waterman's wrists, then ankles, and used a recording electrode farther up his arms, then legs, to measure the nerve conduction speeds. Waterman's body blindness lifts on occasion, specifically to pain and temperature. "These tests are usually slightly unpleasant rather than painful," Cole told me, "but whatever I did to Ian, I first did on myself—a rule I've stuck by since, no matter what experiment we've done together." Waterman replied, "I love it when doctors say, 'This isn't going to hurt,' but my trust in Jonathan began during those early tests." The neurophysiological examination took days, then weeks, until slowly a picture began to emerge. Cole diagnosed Waterman with acute sensory neuropathy syndrome, a condition so rare that it had been described for the first time in the medical literature only three years previously.

The main axons in our limbs are a weave of different types and sizes of nerves. They contain large, fast motor fibers through which signals shoot from the brain at around 180 kilometers (110 miles) per hour, ordering muscles to contract. In contrast, sensory fibers conduct in the opposite direction, bringing information of the outside world inside and to the brain. The smallest and slowest of these—the A-deltas and C-fibers—come from the temperature, pleasure, and pain sensors that we encountered with the vampire bat. The somewhat larger and faster A-beta fibers convey messages from the touch cells discussed in relation to the star-nosed mole. Finally, the largest sensory fibers—the A-alphas—fire at the same speeds as the motor fibers, coming from proprioceptors such as our muscle spindles. Cole's tests revealed that Waterman had suffered a comprehensive loss of certain sensory nerves. "We discovered that the small A-deltas and C-fibers were intact, which is why Ian still senses heat, cold, and pain. However, below his neck, he had lost all the large fibers—the A-alphas and A-betas—explaining his lack of proprioception, as well as touch." In an unusual complication, the virus had inflamed and destroyed these nerves, thereby sealing Waterman's fate. His proprioceptors were spared, but because their signal never reaches the brain, they were silenced. Cole's tests also showed that Waterman's motor nerves were working normally. So how was his brain commanding the movement of limbs without knowing their whereabouts?

When in the hospital, Waterman had often been asked by doctors to perform the standard neurological test of bringing his finger to his nose. "I didn't realize I couldn't do this until they asked me," he said. "The trouble was that my forefinger was just as likely to go up the nurse's nose or in the doctor's eye as onto my nose." He realized that his wandering arms and ungovernable digits were proof that he had lost not movement but control of it. On arriving back home, Waterman resolved to regain mastery over his body. He started simply: "I concentrated on how to feed myself: on moving my arm, bending the elbow, then the wrist, clenching my fingers tightly around what I wanted to eat." He teased each action apart into smaller and smaller components. Then, with grim determination, he watched his body reassemble these parts. Slowly but surely, he found he was able to get his arm to respond. "But then my other arm would lift off the bed and float aimlessly. Why? It had always looked after itself before. Why couldn't it now?" It seemed that, a little like Inky, Waterman's arms tended to roam whenever his gaze was turned.

Lying on his back and staring up at the same patch of ceiling, Waterman began to wonder if he could command his body to sit up. "Having thought through the whole process and deconstructed it into simple movements, I looked down at my chin," he told me. "I folded and tucked it to my chest, as if to start the first part of the curl." Then, eyes trained on his stomach, he seemed able to tighten its muscles and peel himself off the bed. "I was close, but couldn't quite make it. What was holding me back? My arms, my bloody arms, were in the way.

Their weight was holding me down." So looking at his body again, he painstakingly repositioned it and tried once more with renewed focus. "This time, as I curled up, I swung my arms forward. Then I did it. Did it! When I got there I was so damned euphoric, I nearly fell out of bed."

As Waterman began to take back control of his body, the importance of combining his conscious attention with vision dawned on him. Not only did his eyes enable him to see the world but also they would be the means by which he would propel himself through it. By staring at a limb, a digit, any joint, once again his brain was able to issue commands and move it. Vision and intention could stand in for his loss of proprioception. Months, then years of intense daily observation and grueling rehabilitation followed as he dissected and relearned the motor skills to sit, stand, turn, bend, and walk.

Waterman would become the first person in history to study commonplace human movements in such forensic detail. To this day, he remains utterly reliant on his eyes. "I control all my movement with my sight," he said. "If I look away from my hand, I lose it and my hold on it." This connection is always tenuous because it relies on things beyond Waterman's control. If the lights fail, he falls. If blinded momentarily by bright fireworks or a sneeze, he falls. If his focus falters for even the briefest of moments, he falls. Furthermore, the immense effort of conscious attention he must bring to bear shows how much we take this sense for granted.

Our brain receives an unending flow of information from the many thousands of our proprioceptors around our body, all of which it processes without our awareness. "It would be a waste of mental effort if we had to decode consciously the barrage of activity continuously transmitted from our muscle spindles," Cole explained. Mercifully, we remain oblivious to the position of individual spindles and the degree to which they stretch. Cole continued, "When we command a movement, we do not have to think how to move all the muscles to make it happen." Charles Sherrington long ago recognized that we are aware of the aim rather than the many minute acts required en route, allowing most motion to unfold without our awareness. If our body is the puppet, then proprioception is our internal puppeteer, orchestrating movement on our behalf. Robbed of this sense, Waterman must view and command his body from the outside. He has become both puppeteer and puppet; his eyes enable his brain to pull the strings so long as he has focus. "Ian cannot think of much else when moving as it entails so much effort," said Cole, "so vision is a poor substitute for his loss of proprioception." Waterman added, "You might remember a walk for the views; I remember it for the walking."

Today, Waterman can fool most people. Although he is now in his seventies and has resorted reluctantly to using a wheelchair, one feels it neither restricts nor constrains him. He fidgets, he swivels, he gestures with his hands as he speaks; he gives the distinct impression that at any point he might nimbly get to his feet. Meeting Waterman, one has to remind oneself continually that he is performing and that this exacts a huge toll. "I didn't realize what I had lost until I met Jonathan. I didn't even know the word *proprioception*," Waterman told me. "Sometimes I wake up in the morning and the knowledge of how much mental effort I'll have to put in to get by makes me feel down. It's like having to do a marathon every day." Even a common cold renders him unable to focus and therefore unable to move. "Ian is like any top athlete striving for peak performance," Cole explained, "but he has to give his best at every moment and without an audience cheering him on." Waterman may describe with honesty and eloquence what Sherrington once deemed "indescribable," but his daily marathon reveals the profound purpose of our sense of body.

It was perhaps inevitable that Cole would introduce Waterman to his mentor. Oliver Sacks had once, years before, encountered the condition in a patient of whom he said at the time, "so far as I know, the first of her kind, the first 'disembodied' human being." On listening to Waterman's story, Sacks declared it "at once terrifying and inspiring." He was struck by the "almost superhuman resource and will" that was required to live without "the most elemental human, or animal, sensation." One might presume that this sense of body, lost to Waterman, reaches its apogee in our eight-legged Proteus. When I told him that of all humanity, he is the one with whom the octopus has most in common, he was delighted. Instantly, he quipped, "Call me Inky."

*

As Pasquale Graziadei finely dissected the octopus arm and looked at the proprioceptors within its muscle, he teased apart their nerve network. At first glance, the nerves appeared to lead no farther than the arm. This arrangement had never been seen before in any other creature. Subsequent anatomical investigations by colleagues confirmed that the main nerve route, from an octopus's arms to its head, is much reduced, connected by only relatively few nerve fibers.

Today, scientists wonder whether any proprioceptive information whatsoever gets to the octopus brain. Consequently, some describe the octopus's arms as being "curiously divorced" from its brain. Given they make up most of its weight, this implies that much like Ian Waterman, an octopus is estranged from its body. The late Martin Wells, an authority on cephalopods, once wrote, "One can only approximate to the octopus condition by using some complex piece of machinery (or another animal) as an extension of one's body." In this framework, the octopus becomes an amalgam of driver and car. When we accelerate a car, we do not feel the quickening of machine wheels; similarly, when we spur a horse to gallop, we do not feel its legs move. Instead, we judge the success of these commands through senses beyond proprioception, by watching the landscape flit past faster or feeling the wind in our hair. Benny Hochner said, "Even if some proprioceptive information does in fact reach the octopus's brain, the question remains whether the animal uses information from the arm proprioceptors to understand what its arms are doing. I believe it does not." Assuming the octopus had either fingers or a nose, if asked to perform the classic test for proprioception, Hochner told me that it would fail: "The octopus could not use proprioception to 'bring its finger to its nose.'" So how these creatures move with such fluidity and shape-shift with such seeming purpose was a mystery.

Hochner and the Octopus Lab decided to film an everyday octopus action. "It's relatively simple to train an octopus to reach for a target," Hochner said. They lowered a green disc into the tank and rewarded the animal that touched it with a treat. "The Naples lab found anchovies worked well, but we used shrimps. Actually, octopuses are so curious they often don't need the bribe to reach out their arm." The scientists set up twin cameras at right angles to one another outside an octopus tank. They focused both lenses on the animal's arm as it shot out to full stretch. The movement took no more than 1 second. "We did a kinematic analysis," Hochner said. "We combined the footage from the two aspects to compute the reaching movement in three dimensions." Then they repeated the experiment. Time after time, octopus after octopus, the results revealed different arms at different times unfurled in a surprisingly stereotypical way. Hochner recalled, "We kept seeing the same movement. A wave of stiffness would start near the head and propagate down the arm to its tip, a lot like a children's party blower." Even though an octopus could reach out its arm in infinite ways, it did not.

"This particular reaching movement is always the same and actually very simple. This was our first insight," Hochner said. "The kinematics showed that it has only three degrees of freedom." This term, from robotics, describes the number of axes in which a machine arm can move. "The first two degrees of freedom of the octopus arm reach are at the base of the arm dictating the roll and pitch of where the arm aims in three dimensions. The third controls the wave propagating along the arm." In humans, our joints restrict this number; whereas our shoulder has pitch, yaw, and roll, as does our wrist, our elbow has only pitch. "Our arm is said to possess seven degrees of freedom," explained Hochner. "It can reach into any point in space with just six degrees of freedom. The seventh is used to overcome obstacles encountered. But this is still double that of the octopus's arm reach." Despite the octopus's soft-bodied flexibility, it does not have, as often assumed, endless possibility. Its basic, pared-down movements are doubtless easier to master. The mystery shifted from how, without proprioceptive input to the brain, the octopus controls complex movement to how it controls any at all.

Although the octopus boasts the largest brain of any known invertebrate, this organ contains only one-third of its half a billion neurons;
the rest—all 300 million or so—are embedded in its eight arms. Hochner decided to look at the arms apart from the animal in isolation from
its brain. In the wild, octopuses occasionally lose a limb, only to regrow
it entirely. Amputated cephalopod arms had been investigated three
decades previously, with strange and startling results. A scientist had
discovered that when a piece of dried sardine was placed on the tip of a
disembodied arm, it continued to act as if it were a normal day on the
reef. Its suckers grasped the scrap of food, passing it along over hundreds more, like a conveyor belt, and transporting it back to where the
mouth would have been, were it still there. Moreover, when that piece
of food had been infused with noxious chemicals, the same suckers
would promptly reject it. Our arms pick out the best bits on offer at the
buffet, but imagine if they did so after being severed from our body.

Hochner set out to explore whether such independence might extend to locomotion. The team returned to the kinematic setup. This
time, they carefully arranged an amputated arm in the tank. With
cameras running, they stroked a few of its suckers. "We wanted to
see if it would react to natural sensory input," Hochner said. The arm
responded, immediately striking out. The researchers triggered and
filmed more reaches. Hochner added, "An amputated arm continues to
behave normally for about half an hour, so each one allowed us to film
ten good arm reaches." They entered the accumulated footage from
both cameras into the computer and compared their results against
those from the earlier studies of freely behaving subjects. Hochner
recalled, "The amputated arm movements showed the same trajectory
and the same velocity profile—acceleration, peak speed, deceleration—
of a natural arm reach." Their actions were not so much similar as nearly
identical. "We could not even distinguish between the two movements
when they were plotted as graphs." The team had discovered that the
arm unfurls in the same way, whether with or without a head. "This was
our second key insight in this series of experiments. We realized that
the planning, computation, and execution of this movement is embed-

ded within the neuromusculature of the arm itself," said Hochner. "It all happens at the level of the arm without the need for the brain's input." Octopus arms can act of their own accord, without the brain's bidding. They can do so because their repertoire of motions, constrained by few degrees of freedom, is highly simplified. "We scientists are happy when we pose a hypothesis and the result reassures us. This is the reward of playing with science." A cornerstone to this strange autonomy is found in the octopus's similarly strange sense of body.

Although Ian Waterman and Inky are unaware where their respective limbs are and what they are doing at any particular moment, there is a world of difference. Whereas Waterman's arms once wandered randomly when his eyes were turned, those of the octopus seem able to respond to the world with resolve. Whereas Waterman has lost his proprioceptive puppeteer, the octopus has not. The arms of an octopus not only gather the information from their muscles' stretch sensors; they also process it without recourse to the brain. This enables them to monitor and adjust their position by firing off commands to their muscles through motor nerves. "I do not know of any other animal that uses proprioception in its peripheral nervous system this way," Hochner said. "The octopus's unique evolutionary solution to avoid overloading the brain was to offload the bulk of motor control to the arms." Consequently, while one arm sets about winkling a whelk from its shell, another might undertake an exploration of a rocky crevice. Depending on whether this arm encounters an irritable crab or a juicy clam, it will either withdraw or elect to probe further. Meanwhile, the octopus may remain none the wiser to any of these actions. Hochner suspects that due to a ring of nerves that links the arms but bypasses the brain, the arms can send one another information to coordinate themselves into more unified actions. Perhaps they join forces to dig, foray, and fetch. According to Hochner, "They might even be able to recruit each other into complex movements like swimming. The truth is that we simply do not know as it has not yet been studied." Any of these behaviors, simple or sophisticated, can unfold without troubling the brain, under the guidance of their exceptional sense of body.

The Octopus Lab has shown that in certain situations, Inky's brain can take charge. Referring again to the neurological test, Hochner offered a hint: "Although the octopus's central brain cannot give the arm instructions on how to 'reach its nose in the dark,' it could if the lights are turned on." For the first time, the research team has demonstrated that the animal can override the independence of its arms with the sense of sight. In another series of experiments, they observed test animals guiding their arms, though slowly, through a transparent maze in search of a tasty treat by using their eyes. "So, in principle, an octopus could reach for its nose when the lights are on, but in practice it would take a long time, just like it did for those animals in the maze experiment," Hochner said. "This is because although it can follow the tip of its arm with its eyes, it will still involve lots of trial and error to finally find its nose." Once again like Waterman, Inky can use vision to locate and thereby direct its limbs; like Waterman, eyesight seems to draw Inky together, creating coherence and giving it mastery over its body. In all likelihood, this happens during elaborate behaviors, perhaps when giving chase to a plump crayfish or making an escape from a high-security aquarium. Otherwise Inky remains ignorant of its arms' activities and leaves them to their own devices.

*

Peter Godfrey-Smith, of the University of Sydney, is a scuba-diving philosopher with a deep affinity for octopuses. "Many years ago Thomas Nagel used the phrase *what's it like* in an attempt to point us toward the mystery posed by subjective experience," he wrote in his book *Other Minds: The Octopus and the Evolution of Intelligent Life*. He was referring to Nagel's renowned essay that claimed "there is something that it is like to be a bat," "something it is like *for* the organism." Godfrey-Smith raised this question with regard to the octopus: "To do this involves a kind of imaginative leap, an attempt to place ourselves in something like their perspective." Scientists might steer clear of such questions, as did Ron Douglas when asked what it is like to see through the eyes of a deep-sea spookfish, but not so philosophers. Godfrey-Smith admitted freely that this "is not doing science, but it can be guided by science." Like Nagel, he conducts such thought experiments in hopes of shedding light on one of the world's greatest mysteries, perhaps *the* greatest mystery: the "hard problem" of consciousness and how neural matter gives rise to sensation and sentience. "There are two sides to the world that have to fit together somehow, but do not seem to fit together in a way we presently understand," he explained. "One is the existence of sensations and other mental processes that are felt by an agent; the other is the world of biology, chemistry, and physics." The question is: "Does it feel like something to *be* one of these large-brained cephalopods, or are they just biochemical machines for which all is dark inside?"

It is impossible to know with certainty whether another human is conscious or a brain-dead zombie—as the philosopher David Chalmers notoriously declared—let alone another animal. However, Godfrey-Smith argues that octopuses are conscious beings, and he is not alone in such beliefs. On July 7, 2012, when a prominent international group of neuroscientists, neurophysiologists, neuropharmacologists, and neuroanatomists drew up the Cambridge Declaration on Consciousness, our flexible friend was included on their list. So having assumed that there is "something it's like to be an octopus," Godfrey-Smith set out to explore what that might be.

In Harper Lee's *To Kill a Mockingbird*, Atticus Finch famously said, "You never really understand a person until you consider things from his point of view. . . . Until you climb inside of his skin and walk around in it." To get under an octopus's skin and into each of its eight limbs, we must consider its full array of senses. Of all of them, perhaps its proprioceptive sense of body is the most difficult to conceive. As Godfrey-Smith reminds us, unless it resorts to vision, "it is not clear how much awareness an octopus has of the location of its own arms much of the time." However, he goes on to say, "I conjecture (and this step is purely a conjecture) that an octopus uses attention to 'pull itself together' on some occasions, but also, when attentional focusing is absent, the arms are allowed to carry on some local exploration of their own." Some philosophers propose that these two different states of attention might introduce a split in consciousness. They theorize that one subject could become two: one based in the brain, the other in the arm network as a whole. Furthermore, some suggest that given each arm behaves as if it has a mind of its own, perhaps these subjects face further fragmentation: two could become nine. Godfrey-Smith doubts whether individual arms are centers of experience in their own right. He also doubts the existence of a second self. Instead, if octopuses have consciousness, he proposes the existence of a single experiencing self, albeit radically different from consciousness as we know it—in his words: "One that is more or less extensive at different times—one that incorporates more or less of the animal's body." A consciousness that expands when the animal uses vision to "pull itself together," then recedes, leaving the arms through proprioception to take back control of themselves. It is a consciousness that seems to shape-shift like the octopus's body, challenging Wittgenstein's unquestionability of body and dissolving all certainty. Indeed, when Godfrey-Smith tries to imagine what it is like to be an octopus, he concludes, "I find myself in a rather hallucinogenic space, and that is everyday life for an octopus."

*

The common octopus, along with its not-so-common relatives, boasts a unique sense of body. It is not set apart by its sensors—both its and our muscles are studded with receptors that fire when stretched—or by sensory abundance, like nearly all the other creatures described in previous chapters. Rather, the way it processes this sensory information is so distinct that the octopus may have more in common with a person who lost this sense of body. Like Ian Waterman, octopuses cannot turn to proprioception to know the whereabouts of their limbs; they too must rely on vision. But unlike him, their arms remain guided through this sense to operate of their own accord. Benny Hochner attributes such autonomy to the idea that the octopus has a different "embodiment" from any other creature on this planet. The ways in which this word is used reveal much about the mercurial sense of proprioception. When Waterman first fell ill, he said he felt "disembodied." Oliver

Sacks's first case had used the same words on telling him, "I can't feel my body. I feel weird—disembodied." "This was an amazing thing to hear, confounded, confounding," Sacks recalled, but then agreed: "In some sense she is 'pithed,' disembodied, a sort of wraith. She has lost, with her sense of proprioception, the fundamental, organic mooring of identity." It is why he described this sense as "crucial for the perception of one's own body, the position and movement of one's limbs in space, crucial indeed for the perception of their *existence*," making it "arguably more vital than any or all of the other five senses put together"—and arguably more vital than the combination of any or all of the twelve senses discussed in this book. However, today Waterman's outlook is radically altered, and it challenges Sacks's assumptions. "The thing is, I am the most embodied person I know. I have to be aware of all my body, all of the time. I cannot let my mind drift." Jonathan Cole clarified, "Although Ian lacks a proprioceptive sense of embodiment, he is more cognitively embodied than anyone I know as well. He must consciously attend to his body all the time or risk its loss." It is this effort, his daily marathon, that reveals how much of our movement occurs automatically and without our awareness. Cole concluded, "You could say that Ian actually reveals the Inky in each of us. As Inky wields his arms and suckers without recourse to his brain, the reader is probably using their arms and fingers to turn the pages of this book without even noticing."

"The octopus shows that we cannot separate the brain from the body; the brain *is* the body, and in the octopus the body itself contributes to its emergent intelligent behavior," Hochner said. He borrows the term *embodied intelligence* from robotics to emphasize that the body is responsible for some of the cleverness in how an organism interacts with the world. In 350 BCE, when Aristotle waded in the shallows of the Aegean Sea, he was not impressed much by his encounters with octopuses. He even declared in *The History of Animals* that "the octopus is a stupid creature." Yet today our view of this creature has evolved. Peter Godfrey-Smith said, "If we can make *contact* with cephalopods as sentient beings, it is not because of a shared history, not because of kinship, but because evolution built minds twice over. This is probably the closest we will come to meeting an intelligent alien." Discoveries over the last two decades into the octopus's sense of body and its singular embodiment raise profound questions about how we define and classify intelligence. Perhaps now more than ever before, we can grasp just how different the creature's intelligence is to our own. Aristotle could not have known how wrong he was.

The Duck-Billed Platypus:
An Afterword

I N 1799, A PECULIAR DRIED SPECIMEN LANDED ON A DESK AT the British Museum. It had been shipped from the farthest reaches of the empire, a place called New South Wales. The assistant keeper of natural history, Dr. George Shaw, had never set eyes on anything like it. Neither had colleagues, nor, for that matter, anyone beyond the shores of what would soon become known as Australia. Like a mythological chimera, it seemed to fuse features from different creatures: the sleek furred body of a mammal with the webbed feet and shovel-like bill of a duck. Shaw wondered whether he had fallen victim to a colonial prank. "I almost doubted the testimony of my own eyes," he wrote in his *Naturalist's Miscellany*. "It naturally excites the idea of some deceptive preparation by artificial means." Yet try as he might, he could not discover any stitches, any glue or grafting, or any hint of human hand in its creation.

Eventually he decided that the curious specimen was not a hoax but evidence of "a new and singular genus." He named it *Platypus anatinus*, Latin for flatfoot duck. Shaw did not know that the name *Platypus* had already been claimed by a genus of wood-boring beetle, so another scientific epithet had to be conceived. Nonetheless, the original stuck. The duck-billed platypus—*Ornithorhynchus anatinus*—remains one of nature's most unconventional creatures. If anything, over the years it has been found even more curious, so that today, in a group, they are called a paradox of platypuses. Although like other mammals, it nurses babies with milk, it lays leathery eggs like those of reptiles. The males bear spurred claws on their hind legs, charged with snake-like venom. Furthermore, research of its improbable bill revealed a more improbable sense.

As the sun sets behind the gum trees, a platypus emerges from its burrow. Ungainly on land, it slips, streamlined and silent, into a river. As its paddle-shaped feet and beaver-like tail propel it underwater, its eyes, nostrils, and ears seal tight. Without sight, smell, and hearing, it searches for shrimp and crayfish by tracking its bill side to side over the stony riverbed like a treasure hunter with a metal detector. Animals generate weak electric fields when they contract their muscles. The bill is covered in tens of thousands of microscopic electric sensors that can detect these fields, enabling the platypus to home in on its prey with preternatural accuracy.

In 1909, the German biologist Jakob von Uexküll coined the term *Umwelt* to describe the slice of the surrounding environment sensed by an organism. Like the platypus, we have cones and rods in our eyes that detect part of the electromagnetic spectrum and grant vision. Both its inner ear and ours bear fine hair cells that respond to sound waves for hearing. Both its skin and ours hold touch cells and fibers that feel either the contours of the world or the warmth of another's presence. Both its nostrils and tongue and ours hide receptors to either smell or taste molecular cocktails. Perhaps the tuning of these receptors differs, with diverging sensory ranges or sensitivities, but so far, a platypus and a human occupy broadly similar Umwelts. However, we do not possess its electric sense. Electricity exists in the world, but it remains beyond our ken and beyond our sphere of sentience unless at amplitudes sufficient to activate our pain sensors.

"Our brains are tuned to detect a shockingly small fraction of the surrounding reality," said the neuroscientist David Eagleman. "The interesting part is that each organism presumably assumes its Umwelt to be the entire objective reality 'out there.' Why would any of us stop to think that there is more beyond what we can sense?" Therefore, the platypus is a cautionary tale; it reminds us more patently than any of the humans and other animals yet encountered that our experience is not the same thing as reality. We remain insensible to aspects of reality because we can only experience what is first sensed. Uexküll showed that every single environment—from the black watery abyss of the spookfish to the white wind-whistled skies of the great gray owl—offers myriad "realities" to different creatures. To Eagleman, the word *Umwelt* consequently captures "the idea of limited knowledge, of unobtainable information, and of unimagined possibilities." However, we may soon find ourselves in a position to do more than imagine such possibilities.

This century heralds a new dawn for the human senses. Thanks to our brain's remarkable neuroplasticity, researchers are already developing implants to cure blindness by "seeing" with tongues and vibration vests to cure deafness by "feeling" sound. Biohackers believe that all manner of extrasensory perceptions are ours for the taking. We should look to the animal kingdom for inspiration. We see only a ten-trillionth of the electromagnetic spectrum; imagine extending our range to perceive infrared heat like a rattlesnake or see ultraviolet light like a honeybee. Imagine experiencing the taste of a catfish, the touch of a star-nosed mole, or the balance of a cheetah. Imagine opening our mind to Shaw's improbable creature and its still more improbable sense. Skeptics may warn against tampering with our brains, but optimists hail the age of the augmented human. "The human Umwelt has been unlocked," championed Eagleman. "We no longer have to wait for Mother Nature to define us—we now need only ask: How do we want to experience our universe?" What do we want our reality to be? A brave new world of sentience awaits.

ACKNOWLEDGMENTS

I AM INDEBTED TO MANY PEOPLE: NOT LEAST TO IAN WATERMAN, Mark Threadgold, Pam Costa, Eşref Armağan, Joan Eroncel, and Concetta Antico. To the scientists who feature in this book and the many who do not, but still patiently and generously fielded my questions. There are three of note. The first is Oliver Sacks, who appears between these pages alongside some of his patients. Books such as *The Man Who Mistook His Wife for a Hat* deepened my fascination with the fragility and diversity of human experience. The second is Desmond Morris. In *The Naked Ape*, he turned to other animals to better understand ourselves. My interest in zoology has always been to better understand myself. It is a mirror we can hold up to satisfy self-obsession; it offers another perspective on why we humans look, act, and feel as we do. The final guiding light is the man to whom I submitted my first zoology essay. Richard Dawkins introduced me to books such as D'Arcy Thompson's *On Growth and Form* and Robert Axelrod's *The Evolution of Cooperation*, and opened my eyes to the rhyme and reason of the natural world. Some of his ideas lodged in my mind. This is not without irony, given he invented the concept of memes. "If a scientist hears, or reads about, a good idea, he passes it on to his colleagues and students," he wrote in *The Selfish Gene*, and "when you plant a fertile meme in my mind you literally parasitize my brain, turning it into a vehicle for the meme's propagation." His notion of the "anesthetic of familiarity" inspired me to write this book; with any luck, some of its stories have now infiltrated you.

Acknowledgments

To Elizabeth Sheinkman, for believing in me. To George Morley, for taking the gamble; you said we would have fun, you were true to your word. To Peter Borland, for shared enthusiasms in how blind worms see the color blue. To the teams at Picador and Atria, I cannot think of a warmer welcome to the publishing world. To Caterina Leone, for being my other eyes. To Connie, Arthur, and Reggie, who brighten my days and keep my senses spry. And to Dan, who suggested that when it came to thanking him, I should say "with whose help, this book would not have been possible." Please know how far from true this is.

NOTES

Introduction

ix   We humans are often described as sentient beings: Daniel Dennett explored the difficulties in defining sentience in *Kinds of Minds*, as did Marian Stamp Dawkins in her 2006 paper, "Through Animal Eyes: What Behavior Tells Us" (also the source of the Huxley quote). Referring to David Chalmers's famous epithet for consciousness, she wrote, "Sentience is 'the hard problem'" of biology, "because we do not know how sentience arises from brain cells, or how, if at all, brains with sentience work differently from brains without it, we have no real idea what to look for in other species in our search for animal sentience." Henry Marsh did not specifically mention consciousness in this quote from *Do No Harm*, but it is implicit in the wider context. In the penultimate chapter of *Mama's Last Hug*, Frans de Waal suggested three levels of sentience: from sensitivity to experience to consciousness. Dennett again: "Here then is a conservative hypothesis. 'Sentience' comes in every imaginable grade and intensity, from the simplest and most 'robotic,' to the most exquisitely sensitive, 'hyper-reactive' human."

x   However, modern science has proved Aristotle wrong: The neuroscientist Christian Jarrett included the widespread belief of Aristotle's sensorium in his book *Great Myths of the Brain*. To learn more of our less-known senses, listen online to Barry Smith's BBC Radio 4 series *The Uncommon Senses*; his quote is from the short YouTube video "We Have Far More Than Five Senses."

xi   "I cannot pretend I am without fear: Oliver Sacks's last op ed for the *New York Times* was on February 19, 2015, six months before his death.

CHAPTER ONE
The Peacock Mantis Shrimp and Our Sense of Color

2   True to its various names: For a visual introduction to this colorful crustacean and many other stomatopod species, visit the website Roy's List,

belonging to Roy Caldwell, a biologist from the University of California, Berkeley. The first half of the peacock mantis shrimp's scientific name (*Odontodactylus*) refers to its toothlike claw; it shares the second half (*scyllarus*) with the sea monster Scylla who tore up ships in Homer's *Odyssey*.

2 One spring day in 1998: Tyson's story broke in the *Mirror* on April 10, 1998, quoting the aquarium manager, Toby Briant. Furthermore, mantis shrimp do not hesitate to take on creatures many times their size; see one defend its home from an octopus in the National Geographic clip "Watch an Octopus Get Knocked Out by a Shrimp."

2 One scientist at the University of California: Sheila Patek (colleague of Roy Caldwell) did a TED Talk on her adventures in shrimp land. "The strike reaches peak forces of 1,500 newtons—that is 2,500 times the animal's body weight." Imagine the equivalent force scaled up into a human heavyweight's punch. The devastating force has prompted some scientists to wonder how the clubs themselves avoid damage. Patek pointed out that a new limb is built when they molt every few months. However, James Weaver and colleagues found that the peacock mantis shrimp club combines materials in a way that makes it stronger than any known synthetic composite. The shrimp still holds the record for the fastest *feeding* strike in the animal kingdom, but no longer the fastest strike. In 2016, Patek recorded the jaws of a trap-jaw ant springing shut at over 100 kilometers (60 miles) per hour.

3 She was looking at a potent phenomenon: Cavitation rarely occurs in nature because the speeds required are so high. The usual example given to demonstrate its potency is the erosion it causes to the metal propellers of speedboats.

3 Just inland from the Great Barrier Reef: There is a short film of Justin Marshall and his Stomatopod Group in action on their Sensory Neurobiology Group website. He not only explains his fascination with the mantis eye but also shows a photograph from his PhD days taken in Mike Land's lab. Sadly, Land died just before publication. Marshall described him as his scientific hero. I was privileged to speak to him during my research, and his book *Eyes to See* was a riveting introduction to the many eyes in the animal kingdom.

7 In 1994, the late neurologist Oliver Sacks embarked: Sacks recounted his journey and his episode of color blindness in *The Island of the Color-Blind*, and meeting Jonathan I. in "The Case of the Colorblind Painter," in *An Anthropologist on Mars*.

8 Achromatopsia is more than a simple absence of color: Knut Nordby was once called "the most famous rod monochromat in the world." Sadly, he died in 2005, so I relied on his eloquent account in the academic monograph *Night Vision*. The story of why he and other achromatopes are dazzled by sunlight is the subject of the next chapter.

10 We perceive its various wavelengths: The rainbow we see (whose colors we may recall with the phrase "Richard of York gave battle in vain") spans ever-shortening wavelengths from red to violet. The ultraviolet sight of

birds, bees, butterflies, and the mantis shrimp means they see wavelengths shorter than violet. At the other end of our visible spectrum, some creatures perceive wavelengths longer than visible red (infrared), such as snakes and, as Chapter 5 reveals, vampire bats.

11 the most conspicuous creature on the Great Barrier Reef: It might be tempting to assume that colorful creatures always have exceptional color vision—the peacock mantis shrimp and most butterflies are cases in point—but color signaling in the natural world is just as likely between different species. Beetles, for example, display an inordinate array of startling colors, many of which are invisible to other beetles, but serve as "don't eat me, I'm toxic" warnings to their avian predators.

11 Moreover, we now know that both are saturated with: The first opsin discovery came in 1876, but scientists have since sequenced more than 1,000 different opsins in animals from jellyfish to humans. Humans have four for vision: three color opsins and one called rhodopsin. Debate still surrounds the date of the oldest light-sensing opsin and the start of vision, but we now know that our opsins share a deep past with those of invertebrates such as the mantis shrimp, and our color opsins predate rhodopsin. This means our color vision appears older than the visual sense of Chapter 2. We also have other opsins that sense light for reasons beyond sight, as the one from Chapter 10.

13 it pales when compared to that of Tyson: The natural world is full of creatures with more color receptor types than we have. Reptiles and some freshwater fish have four, and as Land observed earlier in the chapter, "Some birds and butterflies have as many as five."

13 "the richest, most harmonious chorus of colors: The quote is from an episode of *Radio Lab*, "Color," May 21, 2012. "A thermonuclear bomb of light and beauty" is how the online comic strip *Oatmeal* describes shrimp sight, adding that when combined with its knockout punch, it makes the animal a "harbinger of blood-soaked rainbows," https://theoatmeal.com/comics/mantis_shrimp.

14 It is often called Daltonism in memory of the first person: In a talk he gave to the Manchester Literary and Philosophical Society on October 31, 1794, John Dalton said, "That part of the image which others call red, appears to me little more than a shade or defect of light; after that, the orange, yellow, and green seem one color, which descends pretty uniformly from an intense to a rare yellow, making what I should call different shades of yellow." On his death in 1844, his eye was preserved; in 1995, John Mollon extracted DNA and discovered that Dalton was a dichromat missing green cones.

15 Gabriele Jordan had grown intrigued by de Vries's long-forgotten paper: She had been directed to it by her supervisor, John Mollon. Jordan also told me that Mollon wondered whether "the anomalous father was actually de Vries himself."

19 On the other side of the world, on the east coast of Australia, another woman: Concetta Antico can be heard online at BBC World Service, *Outlook*, "The artist who can see 100 million colors," January 2015. Her artwork can be seen online at the Concetta Antico Gallery.

CHAPTER TWO
The Spookfish and Our Dark Vision

24 On a calm and cloudless day in July 2007: Ron Douglas and Julian Partridge made a broadcast for BBC Radio 4 *Nature* on board the FS *Sonne*, from which some of their quotes are sourced. You can find it online: "Life in the Trenches," October 29, 2007.

24 The Bathysphere was the most rudimentary of submersibles: William Beebe was later called the "Cousteau of his generation." He and Otis Barton undertook several deep-sea descents from 1930; most broke world records but their thirty-second dive on August 15, 1934, was by far the deepest.

25 "it was no longer scarlet but a deep velvety black": The producers of the movie *Jaws* misled us: even a few meters underwater, blood looks black. Beebe noted an oddity as the rainbow's red, orange, yellow, then green disappeared: "It is strange that as the blue goes, it is not replaced by violet—the end of the visible spectrum. That has apparently already been absorbed." We now know that seawater is most transparent to blue light. However, on the record dive, they saw a full rainbow in the pitch-black ocean at 750 meters (2,400 feet), albeit briefly: "a strange quartet of fish" colored nearly every hue of the rainbow; "their colors could never have been visible" without their torch beams.

27 It uses the same concept developed by the US Navy: The declassified papers from the US National Defense Research Committee (NDRC) on Project Yehudi are available online: "Part III: Aircraft Camouflage," in *Visibility Studies and Some Applications in the Field of Camouflage*, volume 2 of the 1946 summary technical report of Division 16 of the NDRC.

28 Footage taken at the start of the new millennium: The remotely operated vehicles of the Monterey Bay Aquarium Research Institute captured *Macropinna* three times on camera at depths between 600 and 800 meters (2,000 and 2,600 feet), in or near Monterey Bay. Some of the extraordinary footage can be seen on the institute website.

29 Within its mesh lay a creature: Doli was discovered on July 14, 2007, in a net deployed at 11:00 a.m. and recovered at 3:00 p.m. Douglas told me, "You will frequently catch things you have never seen before. It is possible someone else has, but no one individual can hope to see everything. So it was with Doli. The ten biologists on the expedition combined had spent many years at sea. Yet none of us had ever seen *Dolichopteryx* before." Within the hour, they knew it was special.

31 tales of the inimitable four-eyed spook hit newspapers: Headlines declaring the discovery of the fish (also known as the brownsnout spookfish) on January 8, 2009, included: "Fish with four eyes can see through the deep sea gloom," *The Times* (London); "Four-eyed brownsnout spookfish comes out of the deep," *The Australian*; and "This fish has the world's strangest eyes," NBC.

31 Satellite imagery of the moonlit parts of our planet twinkling: See the NASA video (December 2012) "Earth at Night from Space" on YouTube, made from images captured by the Suomi National Polar-Orbiting Partnership satellite over nine days in April and thirteen days in October.

32 His eyes were able to see in light conditions a billion times dimmer: Simon Ings wrote, in *The Eye*, that a healthy human eye is sensitive enough that, in good conditions, we can see the flame of a single candle burning 27 kilometers (17 miles) away. He further argued that our night sight is comparable to that of cats, foxes, and owls, but "it is our relatively poor hearing and sense of smell that lets us down." The later chapters on the great gray owl and the bloodhound aim to challenge whether our hearing and smell do indeed "let us down."

32 The larger, bulbous ones tend to be the cone cells: This is true in general, but as Andrew Stockman told me, actually "the cones in the center where they are densely packed are as small as rods and not 'cone-shaped.' " Stockman's Color & Vision Research Laboratory offers a meticulously detailed online database.

34 Knut Nordby was the achromatope who accompanied Oliver Sacks: Babies too might experience the world much as Nordby did. Our rods develop long before our cones, so they too are nearly blind in bright light. As with the previous chapter, I relied on Nordby's "Vision in a Complete Achromat: A Personal Account."

35 His blight is our blessing: However, on dark nights, perhaps his blight conferred on him some blessings too. In *The Island of the Color-Blind*, Sacks wondered whether Nordby was a better stargazer than most others because the ancient astronomer trick of averted vision came naturally. "It is not that he has more rods than usual, but that his 'strategy' for night vision is unencumbered by a fovea," Sacks wrote. "He does not make the same inappropriate foveal fixation, but instantly catches the stars with his rods."

41 When I asked Douglas what it might be like to see through their eyes: He also told me, "The eye doesn't see, in the same way that a camera cannot see. Vision happens in the brain; it is a mental construct. Since I do not even know what the world looks like to you, it would be idle speculation to imagine what it looks like to a fish." Wagner agreed: "Ron is right to refer it to philosophers. But maybe a comparison would help: it's like driving a car with tunnel vision forward but at the same time getting the images of the side mirrors. Since we have a fovea, our visual system has to switch between the two, but I should have liked to show (but failed) that in Doli, its main

retina and diverticular retina project to different parts of the tectum, so the two images would be processed (perceived) simultaneously."

CHAPTER THREE
The Great Gray Owl and Our Sense of Hearing

44 The quietness of the owl's flight is unrivaled: An introduction to Nigel Peake's research into the owl's silent swoop can be heard on the podcast "O is for Owl," the Cambridge Animal Alphabet series, online on the University of Cambridge's SoundCloud.

45 In 1963, Masakazu, or Mark, Konishi attended a lecture: At the time of writing this, I was unable to speak to Mark Konishi due to his ill health; sadly, he has since died. I thank Konishi's friend and long-term collaborator Professor Catherine Carr, now at the University of Maryland. Carr kindly double- and triple-checked Konishi's contribution and the sections describing his work; her input was invaluable. I also relied on two filmed interviews, both available online: an archival interview conducted by the Society of Neuroscience, March 29–30, 2006, and the University of California San Diego's "UCSD Guestbook" with Nick Spitzer on June 28, 2006.

45 Vision researchers saw the rod-dense retinas: Of all the owls, the tawny is thought to have the best dark vision, but when Graham Martin compared its absolute visual threshold with humans, he found that on average, the owls were only slightly more sensitive. His *Nature* paper echoes the conclusions of the previous chapter: "This suggests that retinal mechanisms in both man and owl have reached the ultimate in sensitivity." Martin added, in *Birds by Night*, "We might expect to find individual people with visual sensitivity greater than that of individual tawny owls."

47 We are deaf to: Many creatures rely on sound frequencies above and below our range of hearing. For example, dolphins produce bursts of high-pitched ultrasound to "image" their prey, and beluga whales do so to locate breathing holes in ice sheets. Low-frequency sounds, such as breaking waves, waterfalls, or wind hitting mountain ranges, carry for miles, and migrating birds navigating vast distances (such as the bar-tailed godwit of Chapter 11) use these infrasounds as "acoustic landmarks."

48 the loudest sound in recorded history: The Royal Society 1888 publication "The Eruption of Krakatoa, and Subsequent Phenomena" declared of the account of the chief officer of police in Rodrigues that "it is also the only instance on record of sounds having been heard at anything like so great a distance from the place of their origin" (79). Bhatia's article "The Sound So Loud That It Circled the Earth Four Times" more recently repeated the claim, adding that on August 27, 1883, at 10:02 a.m., "the Earth let out a noise louder than any it has heard since"; it also inspired the Boston/Dublin analogy. The *Norham Castle* logbook quote is from Simon Winchester's book.

48 given permission to spend time in Beranek's Box: According to the "Interview of Leo Beranek by Jack Purcell on February 26, 1989," the structure was made with over 19,000 3-feet-long (1-meter-long) wedges, from "seven railway carloads of fiberglass."

49 He later recounted, "In that silent room: The quote is from Cage's 1958 lecture in Brussels, "Indeterminacy," whereas "There is no such thing as silence" is from *Silence: Lectures and Writings*.

49 As minutes pass, these anechoic explorers: In the Radio Lab broadcast "Hallucinating Sound," March 21, 2008, the presenter, Jad Abumrad—source of "After about twenty minutes, I began . . ." (whereas "I started to hear the blood . . ." is George Michelsen Foy)—revealed how our brain can behave strangely in total silence by creating perceptions of nonexistent sounds. He started to hear a Fleetwood Mac song: "I remember thinking, how'd Fleetwood Mac get in here?" And, "the room is quiet, my head apparently not." Sacks observed something similar in his chapter "Musical Hallucinations" in *Musicophilia*, with many patients who on becoming deaf started hearing music, from Bach concertos to Christmas carols to "musical wallpaper." Sacks called them "release hallucinations," explaining that "the brain, deprived of its usual input" starts "to generate a spontaneous activity of its own."

49 in rooms designed since that are even quieter: At the time of going to press, the quietest room on the planet was Microsoft's anechoic chamber, built in 2015, in Redmond, Washington. It appears in the Guinness Book of Records for recording a sound that was −20.6 dBA.

51 their ear openings are hidden: The Teton Raptor Center has a spectacular video clip on YouTube, "Great Gray Owls: Talons, Ears and Eyes," in which an ecologist carefully holds a great gray female and gently parts the feathers of her ruff to reveal her ears, saying, "If you look really carefully you can almost see the back of her eyeball in her skull."

54 The great gray's disc is as well developed and even bigger: The sound-funneling abilities of the great gray ruff have not been measured, but Rolf Åke Norberg has studied the bird and told me that its facial ruff is "very much bigger" than that of the barn owl and "this larger facial ruff probably collects more sound, meaning the great gray could hear fainter sounds than the barn owl, but we don't yet know much about its middle and inner ears." This is why Köppl conservatively suggested that "the great gray has an auditory sensitivity *at least as* great as the barn owl."

56 in his memoir, *Touching the Rock*: Hull died in 2015, but his experience was poignantly portrayed in the film *Notes on Blindness* (2016). In the *New York Review of Books*, Oliver Sacks wrote of John Hull's memoir that there had never been "to my knowledge, so minute and fascinating (and frightening) an account" of blindness.

56 the ophthalmologist Emile Jarval: Credit to Harold Gatty, who wrote, in *Finding Your Way*, "Emile Jarval was responsible for introducing the term 'sixth sense.'"

57 A silent, monochrome movie exists of Karl Dallenbach's experiments: The film can be seen online at the Max Planck Institute for the History of Science, Berlin, as VL Library Item lit39549 (Audio/Film). Also the citation (by Michael Supa, Milton Cotzin, and Karl M. Dallenbach) is listed in the references section.

58 Scientists have shown we hear silent objects through the sounds they reflect: Various laboratories have looked at the ability of both blind and sighted people to sense their surroundings with sound, including that of Lawrence Rosenblum, who writes a superb summary of human echolocation in his book *See What I'm Saying.*

CHAPTER FOUR
The Star-Nosed Mole and Our Sense of Touch

64 It is a rather bewitching mammal: The star-nosed mole can be seen online in a *National Geographic* clip "World's Deadliest: Is This the World's Weirdest Looking Killer?"

64 Ken Catania, the world authority: He is also an authority on extreme animals other than the star-nosed mole—from tentacled snakes to naked mole rats—with a desire to understand them as extreme. He once used fake zombie arms to prove a 200-year-old observation that electric eels will jump out of water to attack. But then he dipped his own hand into the tank, and when the eel dutifully performed, he felt a jolt so powerful that his arm jerked away. "It efficiently activated my pain receptors," he said. The Catania Lab website is a repository of slow-motion videos of star-nosed moles sniffing underwater or foraging for worms. Catania's book *Great Adaptations* is another superb source.

66 "These points or papillae are the seat: Lena Borchers kindly translated Dr. Eimer's paper.

67 A macabre collection: The work of Dr. Fukushi (who died in 1956) was photographed for *Life* magazine. More photographs of his collection are on the website of Tattoo Cultr in "Dr. Fukushi Masaichi [*sic*] and Ilse Koch: Two Twisted Tales of Obsession for Tattooed Human Skins."

71 A *Fox Movietone News* archive clip: The 1928 clip titled "How Helen Keller Learned to Talk" is available on YouTube. It is also the source of the Anne Sullivan quotes.

74 In the 1930s at the Montreal Neurological Institute: I first heard of Wilder Penfield's work from the man who reintroduced awake craniotomies to contemporary medicine, Henry Marsh. In *Do No Harm*, he articulated the many oddities of the procedure; what, for example, "the dualist philosopher Descartes, who argued that mind and brain are entirely separate entities . . . would have said if he could have seen my patients looking at their own brains on a video monitor." Archive footage of Penfield's awake craniotomies from the 1920s can be seen in a short film for the Canadian Medical Hall of Fame, available on YouTube.

75 Over the years, as neuroscientists turned their attention to: The shapes of cortical touch maps have been found to vary wildly between species, and they emphasize the features that are most touch-sensitive. So the touch map of the raccoon has enormous front paws, whereas that of the pig has a humongous snout. Some *unculi* can be seen in Rob DeSalle's *Our Senses* and others in the Blakeslees' book, *The Body Has a Mind of Its Own.*

75 In 2004, a man would walk into a laboratory: Eşref Armağan's website is a great place to discover more about him and to see his artwork: https://esrefarmagan.blogspot.com. Alvaro Pascual-Leone discussed the study of Armağan, blindfolding the sighted for research, and more in a talk to the School of Medicine at the University of Wisconsin in 2009, called "Learning about Seeing from the Blind," available on YouTube.

79 "This is what a brain does,": Pascual-Leone's volunteers were experiencing the visual equivalents to the auditory hallucinations that some people experience in anechoic chambers—what Oliver Sacks called "release hallucinations."

CHAPTER FIVE
The Common Vampire Bat and Our Senses of Pleasure and Pain

84 Its saliva holds as many: Biologists describe vampire bats as venomous. We tend to think of venom as a cause of pain or death, but strictly speaking, it is a secretion that has the potential to disrupt another's physiology. Vampire bat saliva contains all sorts of active compounds. Draculin, discovered in the 1990s, has even been trialed in humans as a blood thinner.

84 No wonder such an adept bloodsucker stoked: Contrary to what Bram Stoker would have us believe, the common vampire bat rarely bites humans, and if it did, it would be hard-pressed to exsanguinate one as it only consumes a tablespoon or two of blood at one sitting. The other two vampire bats (the hairy-legged and white-winged species) prefer the veins of birds.

90 As our hands explore an object: Strictly speaking, the fast, fat, and fat-insulated nerves that serve our tactile mechanoreceptors are called A-fibers; these also innervate the sensor from the octopus (Chapter 12) that creates our strange sense of body known as proprioception, whereas the slow, slender, and predominantly uninsulated sensory nerves are called C-fibers. C-fibers serve pleasant touch, pain, and temperature yet also itch, which is described as a sense in its own right and named pruriception. That said, some pain and temperature receptors fire via the fast A-fiber pathway, for immediate "get-out-of-danger" responses.

92 It was not until 1976, when Costa turned eleven: The letter was from Dr. Mahlon Burbank. He was the first to publish in the scientific literature on the condition of inherited erythromelalgia and, therefore, the first to prove that Costa's condition was not in her mind but in her genes.

93 the eminent English neurophysiologist and Nobel laureate: Aristotle (384–322 BCE) classed pain as an emotion. The Muslim philosopher Avicenna (980–1037) subsequently proposed that pain was a sense independent of touch. Sherrington gave this sense a name and was the first to set out how it would work physiologically: how pain sensors lead to pain perception.

94 the "strangest to appear in scientific literature.": Science writer and geneticist Ricki Lewis wrote these words in his blog on April 21, 2016: "No pain and extreme pain from one gene." He continued, "I've included this story in my textbooks for so long that I recently began to wonder if I was perpetuating an urban legend." The study into the Pakistani street performer's family was led by Professor Geoff Woods at Cambridge University, but his colleague James Cox, now at University College London, was exceedingly generous with his time and patience answering my questions.

97 "In reality, they are lovely, affectionate animals: Jerry Wilkinson visited Schmidt's captive colony and saw behaviors that reaffirmed his conclusions with wild bats. He told me, "One of Uwe's grad students took me to see the captive bat colony. He called one of the bats by name, and it hopped to the front of the cage, allowing the student to scratch it around the head, very much like a dog would come to its owner for petting. While anecdotal, that is probably the best evidence I have that vampire bats enjoy being groomed."

97 "It had been shown only once before in vertebrates,": Schmidt is referring to research that had carefully blindfolded pit vipers and then encouraged them to strike a heat source (the tip of a soldering iron). The snakes had targeted heat with such accuracy that the scientists E. A. Newman and P. A. Hartline concluded, "Very impressive, and for a mouse deadly."

98 "To begin with, we thought we would find the heat receptors: Although Newman and Hartline claimed snakes have infrared "vision," this sense does not rely on their eyes but on their pit organs. Strictly speaking, their unusual sense was, like the bats, part of the somatosensory system.

100 Moreover, we know now the TRPV1 sensor is but one: For example, David Julius and his team have found a different heat sensor in the pit organs of snakes that accounts for their thermoperception; it is called TRPA1.

103 This Aristotelian concept remains widely accepted: Although Charles Sherrington coined *nociception* long before the discovery of hedonoceptors (in his 1906 magnum opus, *The Integrative Action of the Nervous System*), perhaps he would have been interested to learn that pain and pleasure share the same slow touch system and can influence one another. McGlone told me, "We all know how rubbing a pain can make it feel better. New research shows how activating hedonoceptors, through a gentle stroke, can reduce the activation of nociceptors and the perception of pain." Studies have shown how stroking the skin of adults before briefly touching them with a burning heat means they both sense and perceive less pain. Similarly, brain scans of babies undergoing heel pricks also showed a reduced pain response

if the heel is first stroked. This growing body of evidence suggests that pleasure and pain can be considered two sides of the same sense.

CHAPTER SIX

The Goliath Catfish and Our Sense of Taste

107 Of all the predators in the Amazon: In 2009, Jorge Masullo de Aguiar reputedly caught the largest Goliath in the Amazon blackwaters. In the trophy
photograph (which can be seen online), the fisherman stands alongside his
catch, dwarfed by its 2 meters (6.5 feet).

107 All fish possess a lateral line: A series of specialized mechanoreceptors
(called neuromasts), arranged along head and flanks, grant the fish a tactile
sense to feel the pressure waves of another creature moving up ahead, the
flow of a river, or the currents of an ocean.

107 The longest recorded human tongue: According to Guinness World Records, the longest human tongue belongs to an American, Nick Stoeberl, and
measures 10.1 centimeters (nearly 4 inches).

107 "The catfish is the iconic animal when it comes to taste: In 1977, the sensory
biologist John Caprio conducted a classic study of the absolute sensitivity
of a catfish's taste nerves and discovered that catfish have the lowest taste
threshold among vertebrates. He told me a twist to the idea that the catfish is
a swimming tongue: "It's as if the tip of your tongue grew out of your mouth
and covered your entire body."

108 Inspired by the American polymath: To learn more about the late, great neurologist Charles Judson Herrick, George W. Bartelmez's biographical memoir
is a good start and it can be found online at the National Academy of Sciences.

109 More recently, scientists found glutamate or umami receptors: This happened
in 1996, even though the taste channel was proposed nearly a century earlier
by the Japanese professor Kikunae Ikeda. Scientists are still looking for more
basic tastes; most recently, channels for fat and for calcium were put forward.

109 Yet most of a food's flavor: Bartoshuk and others have proposed that flavor
might be considered a sense in its own right. In his 1826 book, the famous
French gourmand Jean Anthelme Brillat-Savarin declared, "For myself, I am
not only convinced that there is no full act of tasting without the participation
of the sense of smell, but I am also tempted to believe that smell and taste
form a single sense." In *Neurogastronomy*, Gordon Shepherd added, "Brillat-
Savarin thus identified the important role of smell in taste, but unfortunately
didn't differentiate clearly between taste as a single sense, and 'taste' as a
combined sense of smell and taste. This second combined sense is 'flavor.'"

109 In the case of flavor, our brain hoodwinks us: Gordon Shepherd also said
that "the sense of flavor produced is a mirage; it appears to come from
the mouth, where the food is located, but the smell part, of course, arises
from the smell pathway." According to Bartoshuk, this is also why Aristotle

missed it; when he "bit into an apple, the flavor of the apple was perceptually localised to his mouth so he called it 'taste.'" However, although flavor may be more about smell than taste, it is also about sight, sound, and the way foods feel, from texture to whether spices are activating our heat sensors. It is truly mulitsensory. To further confound divisions between tongues and noses, taste, smell, and flavor, in 2019 Bilal Malik and others found that our tongues can detect odors. They do not argue that you smell with your tongue but that aromas might tweak taste. The work was reputedly inspired when the twelve-year-old son of one of the paper's authors asked why snakes flicked out their tongues to "smell" the air around them.

109 An experiment entailed swabbing the tongues: Jacob Steiner conducted the baby study, and his paper includes photographs of the babies making—often comical—expressions as they were subjected to various tastes.

110 So it is perhaps not surprising that a recent study: Subjects were asked to read lines like, "She looked at him sweetly," and its literal paraphrase, "She looked at him kindly," while their brains were scanned. F. M. M. Citron and A. E. Goldberg found that metaphorical expressions are more emotionally evocative; unlike their literal counterparts, they activate parts of the brain that process emotion such as the amygdala.

110 Taste is as hardwired as our sense of pain: We are born with our five channels of taste fixed (though we can develop taste aversions from eating food that has passed its sell-by date), whereas we learn our smell preferences through life.

115 Dr. Raymond Fowler noticed: Sadly, the American psychologist died in 2015; Bartoshuk kindly told me his story and directed me to Erica Goode's article for the *New York Times* that featured Dr. Fowler.

117 "I called this little fish *piscunculus*.": In fact, Finger first called it (in his 1976 scientific report) *Icthyunculus*, but then decided against "this mishmash of Latin and Greek."

122 They were first discovered scattered: Mary Whitear first described these solitary cells on fish flanks in 1992. Finger et al. (2003) found them in the respiratory tract; Höfer et al. (1996) found them in the digestive tract.

CHAPTER SEVEN
The Bloodhound and Our Sense of Smell

127 Little wonder that trained cadaver dogs: In *Being a Dog*, Alexandra Horowitz, who runs the Dog Cognition Lab at Barnard College in New York, recounted many other stories attesting to the power of a dog's nose: from dogs trained to find drowned bodies—"the odor of decomposition rises to the surface"—to pets behaving oddly when their owners develop cancer. Scientists wonder whether malignant tumors emit "a kind of signature scent" undetectable to our nose. Studies have looked at whether dogs can sniff out melanomas as well as bladder, prostate, breast, and lung cancers.

128 Settles found a willing subject: Bailey belonged to Settles's colleague Lori Dreibelbis, who kindly wrote and told me, "Bailey was an amazing dog who lived to please. She loved going into work with me. She was easy to teach. Her hard work contributed to the education of a PhD student and several scientific publications."

130 Whatever the precise quantities: The exact numbers of olfactory neurons may be debated, but they are always substantially higher in dogs. For example, the psychologist Rachel Herz proposed, "We have approximately 20 million olfactory receptors covering the epithelium of both our right and left nostrils," which looks "paltry to a dog like the bloodhound who has about 220 million." Whereas the psychologist Stanley Coren claimed that the bloodhound "checks in with around 300 million scent receptors." Also of importance, dogs—unlike humans but like, say, snakes, salamanders, elephants, and turtles—have a second surface of olfactory epithelia in their noses, at the bottom of the nasal cavity above the roof of the mouth: the vomeronasal sensory epithelium of the vomeronasal organ. This organ contains specialized receptors that detect airborne messages called pheromones (the subject of the next chapter).

131 Humans do not have a recess, so air whistles through: According to a pioneering study by David Laing, a typical human sniff lasts 1.6 seconds, is 500 cm^3, and reaches 27 liters per minute.

134 The year before his death, Broca wrote in the *Revue d'Anthropologie*: The translation is by John McGann, from his paper "Poor Human Olfaction Is a 19th-Century Myth."

134 "The organic sublimation of the sense of smell: Freud addressed these words to the members of the Vienna Psychoanalytical Society in 1909.

135 She called smell the "fallen angel,": In *The World I Live In*, Keller wrote, "For some inexplicable reason, the sense of smell does not hold the high position it deserves amongst its sisters. There is something of the fallen angel about it."

137 "Does that sound like across-the-board, doggy nose superiority to you?": Avery Gilbert explored this debate on his blog, *First Nerve*. See his post "Stepping in It: How Journalists Perpetuate the Myth That Dogs' Sense of Smell Is Superior to Our Own," July 21, 2015.

138 In 2017, a paper titled "Poor Human Olfaction Is a 19th-Century Myth": In *What the Nose Knows*, Avery Gilbert proposed an intriguing theory to explain why Sigmund Freud perpetuated Broca's microsmaty myth: "The repeated insults of cocaine, nose surgery, influenza, sinus infection, cigar smoking, and finally aging left him with a clinically impaired sense of smell." So Freud's sense of smell was neither heightened (hyperosmia) nor lacking (anosmia), but perhaps instead reduced (hyposmia).

140 "Most people were game: Porter's quotes are from Josie Glausiusz's article "Raw Data."

140 Then they were asked to sniff out: In the same way that Settles focused on the dog's inspiration, other scientists are looking at ours. Noam Sobel (also

involved in the Berkeley study) has scanned the brains of people as they sniff. He argues that our sniff is more than just odor collection, but a vital part of our smell perception, saying it is even "sufficient to generate an olfactory percept of some sort even in the absence of odor."

141 Perhaps that world is available to anyone: In her book, Horowitz noted something similar: "Clearly it's not that we *can't* smell; it's that we largely *don't*," she explained. "I had not bothered to open my mind to the smells [but] the great pleasure in having spent the last years thinking about smelling is that my world has changed color." For example, "Each person in my world has a 'smell face' as my great colleague Dr. Oliver Sacks once described it." Dogs, she concluded, are the "quiet distillers of a world that we have stood up from and forgotten."

CHAPTER EIGHT
The Giant Peacock of the Night and Our Sense of Desire

145 The giant peacock of the night is also known: The American entomologist William T. M. Forbes realized that Aristotle's description of the large silkworm in his *Historia Animalium*, book V, section 19, belonged to *Saturnia pyri*, which is why some now call the giant peacock moth's caterpillar "Aristotle's silkworm."

145 He immortalized it in paint: Vincent van Gogh's painting and drawing— both titled *Giant Peacock Moth*, 1889—are available online at the van Gogh Museum website.

145 reports exist of one smitten Saturniid: Although claims are made for greater distances, according to Professor Kaissling, the only proven report is Rau and Rau (1929) of 5 kilometers (3 miles).

146 The same year, two colleagues, Peter Karlson and Martin Lüscher, coined: Also that year, Karlson and Butenandt wrote, "Having consulted a few colleagues with experience in the same field, we should like to propose to name such substances 'pheromones' . . . to designate substances that are secreted by an animal to the outside and cause a specific reaction in a receiving individual of the same species."

148 Pheromones hint at a dark side to the sense of smell: As the notes to the previous chapter mention, pheromones are detected not by the olfactory epithelium but by a separate nasal structure, the vomeronasal organ. There has been much debate about whether humans have a working vomeronasal organ, but the scientific consensus is that we do not. A comprehensive survey can be found in Martin Witt and Thomas Hummel's 2006 work. That said, Ivan Rodriguez in 2000 identified a possible pheromone receptor in our olfactory mucosa, so although we do not have a specialized organ for detecting pheromones, perhaps we have a special class of chemoreceptors tuned to detect them.

148 "*Pheromone* is a very powerful word: Tristram Wyatt's TED Talk, "The Smelly Mystery of the Human Pheromone," May 14, 2014, is available online, and his textbook is a fount of pheromonal knowledge.

148 The journal *Science* deemed the existence: A 2005 special edition, titled "What Don't We Know?", set out the hundred most pressing questions across the sciences. As well as, What is the biological basis of consciousness? Is ours the only universe? What is the structure of water?—it asked "Do pheromones influence human behavior?" adding, "Identifying them will be key to assessing their sway on our social lives."

149 In 1572, the Princess of Condé: This tale and its telling by Charles Féré is from David Michael Stoddart's book *The Scented Ape*, as are other stories, including Huysmans's "love apples" and the fascinating biology of how, where, and why we sweat.

152 The most famous experiment took place in a dentist's: The first two androgen studies mentioned—by Michael Kirk-Smith and David Booth, then Bettina Pause—looked at androstenone. The third study, by J. J. Cowley and B. W. Brooksbank, considered androstenol. Tamsin Saxton's speed-dating study relied on androstadienone.

156 pheromones come in various guises: Charles Wysocki and George Preti have suggested the four varieties of mammalian pheromones, though Tristram Wyatt remains unconvinced by signaler or modulator pheromones.

158 "It's a connoisseur's reaction of delight,": To see the faces of the babies as the putative pheromone was being wafted beneath their noses, watch Wyatt's TED Talk.

163 "Pheromones are not just about sex,": Research has also focused on alarm pheromones and whether we smell fear. This has been long observed in the animal kingdom; for example, Tristram Wyatt noted how Charles Butler wrote "in his beekeeper's manual *The Feminine Monarchie* published in 1623 . . . that an injured bee's 'ranke smell' would attract other angry bees to sting." More recently, Dietland Müller-Schwarze showed that black-tailed deer release alarm pheromones from a gland on their hind legs when startled. Then in 2008, Lilly Mujica-Parodi scanned the brains of subjects as they sniffed sweat collected from people who jumped from a plane at 4,000 meters (13,000 feet). It activated their amygdala (the part of our brain that registers emotion). She told me that this study provided the first neurobiological evidence that humans emit at least one alarm pheromone.

CHAPTER NINE
The Cheetah and Our Sense of Balance

166 Three years later, on a sunny summer's day: The cheetah's world record is recounted in a *National Geographic* article by Roff Smith, "Cheetah Breaks Speed Record—Beats Usain Bolt by Seconds," alongside a video of Sara

breaking the record, both online. The keeper quoted is Cathryn Hilker, who founded the Cincinnati Zoo's Cat Ambassador Program and reared Sara from a pup. The current head keeper, Alicia Sampson, told me that Sara, whose full name was Sahara, "was a true diva who loved running so much." She recalled the race: "That day in June 2012, she ran as fast as she'd ever run. She also had remarkable balance; she was one of our cats who could easily run if the ground was very wet."

166 Half of its mass is muscle: Alan Wilson would show that cheetah muscle can produce a level of power unseen before in land animals. Whereas Usain Bolt, in his record-breaking run, would have generated about 25 watts of energy per kilogram of body weight, the cheetah can fire up 120 watts per kilogram of body weight.

166 "We had only one valid and ratified measurement: Although Craig Sharp conducted the study in 1965, he did not publish until 1997, when he discovered that the speed widely reported at the time—114 kilometers (71 miles) per hour—was based on dubious science. Apparently the researchers had overestimated not only the distance run but also their timing, and subsequent calculations were inaccurate.

168 Oliver Sacks narrated meeting: Losses of balance are more common than one might think. Ron Douglas, of four-eyed spookfish fame, has lost the balance organs of one ear. He has to take great care walking but finds consolation in never suffering from seasickness—"another reason to head off to the high seas" in search of new and strange species.

169 "It is this 'getting-left-behindness' that the inner ear measures.": Brian Day offered another description: "It's a sense of the force of gravity and a knowledge of which way is up that is with us all the time," unless, of course, we find ourselves in space; astronauts often feel dizzy and disoriented in their first few days of microgravity. Neuroscientists such as Alain Berthoz wonder whether a better understanding of what happens to an astronaut's sense of balance might help people suffering with balance problems.

175 Pun aside, it is as valid today despite subsequent discoveries of fossils: The Laetoli footprints remain the oldest footprints we know for hominins, but the date for the start of human bipedalism has been pushed back to 7 million years ago. The Chad fossil Toumaï (*Sahelanthropus tchadensis*) has replaced Lucy as the oldest known human ancestor that could walk.

177 "Our vestibular system does not operate in isolation,": There is another sense, other than vision, that feeds into balance. It is the subject of Chapter 12 on the octopus: a sense of body called proprioception. When Day conducted galvanic stimulation on a man who years ago had permanently lost this sense (Ian Waterman, also of Chapter 12), Day found that the man's vestibular sense was much more sensitive than that of most others. "I'd say Ian's vestibular response was ten times our magnitude," he told me. "It is up-weighted presumably to cope with his lack of proprioception."

177 Consequently, our eyes play a key role: The importance of vision in the sense

of balance has been tragically underscored by flying accidents. Without sight, our inner ear can muddle acceleration and tilt, leading to what pilots call the "heads-up" illusion. On August 23, 2000, a Gulf Air flight leaving Cairo crashed into the Persian Sea soon after takeoff. Subsequent investigation found that the night was pitch-black, so without vision, the pilots likely experienced the plane's acceleration as tilt and steered it down into the sea. All 143 on board died.

179 Barry Seemungal is a neurologist at Charing Cross: Dr. Seemungal discussed his work, including this particular study, in a radio interview for ABC: *The Health Report* with Norman Swan, broadcast on June 23, 2014. He argued that dizziness and balance problems are made worse by our modern environment: "Since we have not evolved that much over the last ten thousand years, we effectively carry with us the baggage of our evolutionary history.... Moving vehicles, flashing lights, cinema, widescreen TVs, airplanes, this is not the caveman world." The modern world is "very challenging for the brain."

CHAPTER TEN

The Trashline Orbweaver and Our Sense of Time

186 Scientists have shown that orbweavers: A *National Geographic* article by Carrie Arnold, "Spiders Listen to Their Webs," reported on the research of Beth Mortimer at Oxford University. Mortimer has studied the sonic properties of spider silk and explained how spiders generate vibrations to get information. Their eight legs mean "they essentially have an ear covering all different directions" and "because spider silk can vibrate at so many frequencies, the spider can sense movements as small as a hundred nanometers."

186 Few students dare to pass through the door into Thomas Jones's office: A photograph of Thomas Jones holding a golden orbweaver is on the East Tennessee State University website, alongside an article, "Researchers Gaining National Attention for Work on Spider Circadian Rhythms," December 12, 2017. Darrell Moore also makes an appearance, as does the fabled unicorn spider.

188 In the balmy Languedoc summer of 1729: The talk to the Royal Academy of Sciences in Paris in 1729 did not specify what plant de Mairan had used. It was six years before the Swedish naturalist Carl Linnaeus published his *Systema Naturae*, so de Mairan would have been hard-pressed to identify his specimen with any accuracy. However, when other botanists repeated the experiment two decades later, they used *Mimosa pudica*, the touch-me-not. So today it is widely assumed that this was de Mairan's heliotrope. The first quote is from the "Observation botanique," the second from his eulogy; both translations are by André Klarsfeld.

190 inspired one of science's most daring self-experiments: Michel Siffre was not the first to isolate in a cave for science. That accolade goes to Nathaniel

Kleitman and Bruce Richardson; in 1938, they spent thirty-two days in a cave in Kentucky. However, Siffre was the first solitary confinement, and he more than doubled the days spent underground. He is quoted either from his book, *Beyond Time,* or from an interview with Joshua Foer. Footage of him being winched out of the cave can be seen on YouTube: "Scientist Michel Siffre Emerges from Cave after Five Months," September 7, 1972.

192　Jürgen Aschoff, a key scientist in this emerging field: When Aschoff started his pioneering bunker experiments, the term *chronobiology* was not yet coined. He died in 1998. His quotes are from his paper "Circadian Rhythms in Man." There are photographs of his bunker in Michael Globig's article "A World Without Day or Night."

194　British Army sergeant Mark Threadgold: I was introduced to Threadgold by Professor Renata Gomes, chief scientific adviser to the charity Blind Veterans UK. Gomes also kindly fielded very many questions on chronobiology with endless grace.

199　"A rose is not necessarily and unqualifiedly a rose: Colin Pittendrigh is of course referring to Gertrude Stein's famous line—"A rose is a rose is a rose is a rose"—from her 1913 poem "Sacred Emily." Like Aschoff, he was one of the founders of chronobiology. Till Roenneberg (who worked with Aschoff, even volunteering for a stint in the bunker) coauthored (with Martha Merrow) a great introduction to the history of the emerging discipline: "Circadian Clocks: The Fall and Rise of Physiology." Roenneberg would go on to make his name with the discovery that the population is made up of people on a time continuum between early-rising larks and late-to-bed owls. Teenagers tend to be owls; women are more likely to be larks. His book *Internal Time* is an introduction to these chronotypes and the social jet lag that can result from fighting your type.

203　The spider clocks could hardly be more different from our own: Darrell Moore and Thomas Jones have now analyzed eighteen spider species and uncovered vast variety. The sheet-weaver *Frontinella communis,* the bowl and doily spider, has set yet another world record, with a body clock that runs over twenty-nine hours. Recently Jones asked Moore to test his most venomous spider. "We have just discovered that the body clocks of black widows are stranger still," Moore told me. "At first glance their actograms seemed to indicate arrhythmicity, as if they have no built-in sense of time at all. But we now know that unlike any of the other spiders, the widows operate in short bursts of activity."

CHAPTER ELEVEN
The Bar-Tailed Godwit and Our Sense of Direction

206　According to legend: Keith Woodley is the manager of New Zealand's Pūkorokoro Miranda Shorebird Centre. And so his book *Godwits: Long Haul*

Champions is an intimate introduction to the godwit. He points out that the Maori also believe that as "the *wairua*, or souls of the departed, set out on their journey back to *Hawaiki*," their ancestral home, if you listen carefully, you can hear soft murmuring and fluttering sounds "like a flock of kuaka."

208 The bar-tailed godwit still holds the record: In our last email exchange, Bob Gill had breaking news: a reminder that although COVID had stopped international travel for humans, not so for the bar-tailed godwit. "Better get this to press," he told me, "as a godwit tagged this year and followed between Alaska and New Zealand has eclipsed E7's distance." The bar-tailed godwit still holds the record, just not E7.

209 Darwin's paper cited as evidence: Credit goes to Robin Baker and his book *Human Navigation and the Sixth Sense* for alerting me to Darwin's interest in this expedition (and spotting Darwin's misspelling of Baron Ferdinand von Wrangel).

210 European starlings were seen looking to the sun: Emlen's discovery of the birds' celestial compass followed five years after the German ornithologist Gustav Kramer's discovery of their sun compass.

214 As E7 set off on her long-haul flight over the Pacific Ocean: Birds navigate using many senses other than magnetoreception—such as smell, hearing (particularly infrasound), and vision, not simply to calibrate celestial compasses.

216 One laboratory even manipulated millions of these bacteria: The efforts of the Korea Institute of Science and Technology, in collaboration with the University of Twente, can be seen on YouTube under the title "Dancing Magnetotactic Bacteria."

217 The argument descends into more detail, quoting and questioning: Some scientists have questioned whether bird beaks do in fact contain magnetite crystals; instead of crystals, researchers saw iron-rich immune cells. Kirschvink wonders whether the original bird beak experiments were flawed: "Fine-scale investigations of biogenic magnetite are susceptible to environmental contamination, particularly in land animals that live in a world of dirt. The beak research used poor techniques." In a clean lab using careful techniques, Kirschvink and colleagues subsequently found unusually elevated levels of magnetite just *behind* their beaks, in their foreheads.

217 "You couldn't make this stuff up": The words are from Tim Birkhead's chapter on the magnetic sense in his book *Bird Sense*.

217 Magnetic sensitivity has been observed in: Kenneth Lohmann et al. reported it in sea turtles, Nathan Putnam et al. in Pacific salmon, and Alessandro Cresci et al. in glass eels. Eric Warrant et al. reported it in bogong moths and Patrick Guerra et al. in monarch butterflies; John Phillips has seen it in the eastern red-spotted newt, Kenneth Lohmann again in the western Atlantic spiny lobster, and F. A. Brown et al. in the tidal mud snail *Nassarius*; Sabine Begall et al. suggested magnetic alignment in grazing cattle and deer,

Tali Kimchi and Joseph Terkel in the burrowing of the blind mole rat, and Vlastimil Hart et al. argued that dogs too are sensitive to small variations of the earth's magnetic field.

219 "When we compare the senses of time and magnetic direction: Baker added, "The only major difference between the senses of time and magnetic direction may concern sleep. Although there may have been natural selection for the sense of time to function while asleep, for all terrestrial vertebrates, such as Man, there can hardly have been selection to maintain a sense of direction of travel while asleep." Casting new light on falling asleep on a long car journey.

221 Kirschvink's Human Magnetic Reception Laboratory: Kirschvink's lab has a superb website, Welcome to the Magnetic Field Laboratory, that also shows pictures of the human Faraday cage (https://maglab.caltech.edu).

223 The reaction to the paper has been mixed: The microwave quote belongs to Thorsten Ritz (recounted by Kelly Servick in "Humans May Sense Earth's Magnetic Field"); the "we'd know by now" quote belongs to Klaus Schulten ("Our Brains Might Sense Earth's Magnetic Field like Birds"). Ritz and Schulten are renowned in the field as the proposers of cryptochromes.

224 Similarly, according to the cognitive scientist Lera Boroditsky: Her studies were on the Kuuk Thaayorre. Her TED Talk "How Language Shapes the Way We Think" recounted their use of cardinal points in everyday conversations, as well as other experiences living among them.

CHAPTER TWELVE
The Common Octopus and Our Sense of Body

228 Octopuses are renowned escape artists: For tales of octopus ingenuity, intelligence, and much more see both Jennifer Mather's *Octopus: The Ocean's Intelligent Invertebrate* and Sy Montgomery's *The Soul of an Octopus*.

230 chemoreceptors that enable it to "taste" the world through its skin and suckers: Octopus "taste" is in quotation marks because, as Tom Finger wrote in his 2009 paper under the title "A 'Tasty' Embrace," it is debatable whether the creatures' suckered chemoreception should be considered part of a "taste" system. "The octopus appears to use the tentacle chemoreceptors to orient a food object just as catfish use the taste buds on their barbels to orient to food in their environment. Despite these similarities between octopus and catfish taste systems, [they] have evolved separately. Similarities are due to convergence, not common origin."

231 Humans possess some 20,000 spindles: Our bodies have proprioceptors beyond muscle spindles. The ligaments and tendons that attach muscles to bones also have proprioceptors.

232 we can guide our finger to the tip of our nose: This action to test propriocep-

tion is often used by doctors and neurologists, as well as by police officers threatening traffic tickets. The sense is quickly addled by alcohol, so an inability to perform this simple task is a reliable indication of inebriation.

232 As Oliver Sacks noted, the philosopher Ludwig Wittgenstein: Sacks explored the sense of proprioception in a chapter in *The Man Who Mistook His Wife for a Hat* called "The Disembodied Lady." He wrote of Wittgenstein that the thought of losing the certainty of body, of losing the sense of proprioception, "seems to haunt his last book like a nightmare."

232 this is precisely what befell a nineteen-year-old from Portsmouth: Jonathan Cole told Ian Waterman's story in his book *Pride and a Daily Marathon* and its follow-up, *Losing Touch: A Man Without His Body*. The BBC *Horizon* team also filmed them for the 1998 documentary, "The Man Who Lost His Body."

234 Cole diagnosed Waterman with: Acute sensory neuropathy syndrome is so rare that it was not identified until 1980 by a New York neurologist, Herb Schaumburg.

241 Waterman and Inky are unaware where their respective limbs are: The sense that prevents octopuses from tying themselves in knots is in fact "taste." Hochner and colleagues (N. Nesher et al., 2014) have shown that they recognize the taste of their own skin, and this inhibits the suck reflex of their suckers.

241 "The octopus's unique evolutionary solution to avoid overloading the brain: In fact, the Hochner Lab team, Letizia Zullo et al., also looked at the octopus brain and did not find evidence of the classic touch map, seen, for example, in the star-nosed mole, so concluded that the octopus brain neither did nor could monitor where the arms were.

243 the Cambridge Declaration on Consciousness: The declaration was publicly acclaimed at the Francis Crick Memorial Conference on Consciousness in Human and Non-Human Animals, in Cambridge on July 7, 2012, and signed that evening by many prominent scientists, in the presence of Stephen Hawking, at the local Hotel du Vin. They declared: "The absence of a neocortex does not appear to preclude an organism from experiencing affective states. Convergent evidence indicates that non-human animals have the neuroanatomical, neurochemical, and neurophysiological substrates of conscious states along with the capacity to exhibit intentional behaviors. Consequently, the weight of evidence indicates that humans are not unique in possessing the neurological substrates that generate consciousness. Non-human animals, including all mammals and birds, and many other creatures, including octopuses, also possess these neurological substrates." https://fcmconference.org/img/CambridgeDeclarationOnConsciousness.pdf.

243 Some philosophers propose: For example, Carls-Diamante looked to the octopus to suggest "that consciousness is not necessarily unified" and to provide "a counterexample to the claim that a unified consciousness is prerequisite to intelligent behavior."

The Duck-Billed Platypus: An Afterword

248 The assistant keeper of natural history, Dr. George Shaw: Shaw was also a Fellow of the Royal Society and a founding member of the Linnean Society—in short, an experienced naturalist. His monthly installment of *The Naturalist's Miscellany* ran from August 1789 to July 1813.

248 "I almost doubted the testimony of my own eyes,": Shaw continued, "Of all the Mammalia yet known it seems the most extra-ordinary in its conformation; exhibiting the perfect resemblance of the beak of a Duck engrafted on the head of a quadruped."

249 "Our brains are tuned to detect a shockingly small fraction of the surrounding reality,": When the organization Edge asked scientists in 2011, "What scientific concept would improve everybody's cognitive toolkit?" David Eagleman chose the Umwelt.

REFERENCES

Introduction

Aristotle. *De Anima*. Translated by R. D. Hicks. Cambridge: Cambridge University Press, 1907.

Blakemore, Colin. "Rethinking the Senses: Uniting the Philosophy and Neuroscience of Perception." Project of the Arts and Humanities Research Council, July 2014.

Dawkins, Richard. *Unweaving the Rainbow: Science, Delusion and the Appetite for Wonder*. Boston: Houghton Mifflin, 1998.

Dennett, Daniel. *Kinds of Minds: The Origins of Consciousness*. New York: Basic Books, 1997.

Jarrett, Christian. *Great Myths of the Brain*. Hoboken, NJ: Wiley Blackwell, 2015.

Marsh, Henry. *Do No Harm: Stories of Life, Death and Brain Surgery*. London: Weidenfeld & Nicolson, 2014.

Sacks, Oliver. "My Own Life." *New York Times*, February 19, 2015.

Stamp Dawkins, Marian. "Through Animal Eyes: What Behavior Tells Us." *Applied Animal Behaviour Science* 100, no. 1–2 (2006): 4–10.

de Waal, Frans. *Mama's Last Hug: Animal Emotions and What They Teach Us About Ourselves*. New York: Norton, 2019.

CHAPTER ONE
The Peacock Mantis Shrimp and Our Sense of Color

Interviews and personal communications with Sheila Patek, Justin Marshall, Gabriele Jordan, Concetta Antico, and Kimberly Jameson, among others.

Ackerman, Diane. *A Natural History of the Senses*. New York: Vintage Books, 1990.

Brody, J. A., et al. "Hereditary Blindness among Pingelapese People of Eastern Caroline Islands." *Lancet* 1, no. 7659 (1970): 1253–57.

Chiou, T. H., et al. "Circular Polarisation Vision in a Stomatopod Crustacean." *Current Biology* 18 (2008): 429–34.

References

Cronin, T. W., and N. J. Marshall. "A Retina with at Least Ten Spectral Types of Photoreceptor in a Mantis Shrimp." *Nature* 339 (1989): 139–40.

Dalton, John. "Extraordinary Facts relating to the Vision of Colors with Observations." Presented at the Manchester Literary and Philosophical Society, October 31, 1794; published in 1798.

Feuda, R., et al. "Metazoan Opsin Evolution Reveals a Simple Route to Animal Vision." *Proceedings of the National Academy of Sciences* 109, no. 49 (2012): 18868–72.

Hunt, D. M., et al. "The Chemistry of John Dalton's Color Blindness." *Science* 267 (1995): 984–88.

Hussels, I. E., and N. E. Morton. "Pingelap and Mokil Atolls: Achromatopsia." *American Journal of Human Genetics* 24, no. 3 (1972): 304–309.

Jameson, K. A., et al. "The Verdicality of Color: A Case of Potential Human Tetrachromacy." IMBS Technical Report Series, 2014.

Jarrett, Christian. *Great Myths of the Brain*. New York: Wiley Blackwell, 2015.

Jordan, G., and J. D. Mollon. "A Study of Women Heterozygous for Color Deficiencies." *Vision Research* 33, no. 11 (1993): 1495–1508.

———. "Tetrachromacy: The Mysterious Case of Extra-Ordinary Color Vision." *Current Opinion in Behavioral Sciences* 30 (2019): 130–34.

Jordan, G., et al. "The Dimensionality of Color Vision in Carriers of Anomalous Trichromacy." *Journal of Vision* 10, no. 8 (2010): 1–19.

Land, Michael. *Eyes to See: The Astonishing Variety of Vision in Nature*. New York: Oxford University Press, 2018.

Marshall, J., and J. Oberwinkler. "The Colorful World of the Mantis Shrimp." *Nature* 401 (1999): 873–74.

Marshall, N. J. "A Unique Color and Polarisation Vision System in Mantis Shrimps." *Nature* 333 (1988): 557–60.

Neitz, J., et al. "Color Vision in the Dog." *Visual Neuroscience* 3, no. 2 (August 1989): 119–25.

———. "Color Vision: Almost Reason Enough for Having Eyes." *Optics and Photonics News*, January 2001.

Nordby, Knut. "Vision in a Complete Achromat: A Personal Account." In *Night Vision: Basic, Clinical and Applied Aspects*, edited by R. F. Hess, L. T. Sharpe, and K. Nordby. Cambridge: Cambridge University Press, 1990.

Patek, Sheila. "The Shrimp with a Kick." TED video, February 2004.

Patek, S. N., and R. L. Caldwell. "Extreme Impact and Cavitation Forces of a Biological Hammer: Strike Forces of the Peacock Mantis Shrimp *Odontodactylus scyllarus*." *Journal of Experimental Biology* 208 (2005): 3655–64.

Patek, S. N., et al. "Deadly Strike Mechanism of a Mantis Shrimp." *Nature* 428 (204): 819.

———. "Multifunctionality and Mechanical Origins: Ballistic Jaw Propulsion in Trap-Jaw Ants." *Proceedings of the National Academy of Sciences* 103, no. 34 (2016): 12787–92.

Peichi, L., et al. "For Whales and Seals the Ocean Is Not Blue: A Visual Pigment Loss in Marine Mammals." *European Journal of Neuroscience* 13 (2001): 1520–28.

Sacks, Oliver. *The Island of the Color-Blind*. London: Picador, 1986.

———. *An Anthropologist on Mars*. London: Picador, 1995.

Shichida, Y., and T. Matsuyama. "Evolution of Opsins and Phototransduction." *Philosophical Transactions of the Royal Society of London B* 364 (2009): 2881–95.

Silveira, L. C. L., et al. "The Specialization of the Owl Monkey Retina for Night Vision." *Color Research and Application* S26 (2000): S118–22.

Sundin, O. H., et al. "Genetic Basis of Total Colorblindness among the Pingelapese Islanders." *Nature Genetics* 25, no. 3 (2000): 289–93.

Thoen, Hanne, et al. "A Different Form of Color Vision in Mantis Shrimp." *Science* 343 (2014): 411–13.

de Vries, H. I. "The Fundamental Response Curves of Normal and Abnormal Dichromatic and Trichromatic Eyes." *Physica* 14, no. 6 (1948): 367–80.

Weaver, J., et al. "The Stomatopod Dactyl Club: A Formidable Damage-Tolerant Biological Hammer." *Science* 336, no. 6086 (2012): 1275–80.

CHAPTER TWO
The Spookfish and Our Dark Vision

Interviews and personal communications with Ron Douglas, Jochen Wagner, Andrew Stockman, and Alipasha Vaziri, among others.

Beebe, William. *Half Mile Down*. New York: Harcourt, 1934.

Curcio, C. A., et al. "Human Photoreceptor Topography." *Journal of Comparative Neurology* 292 (1990): 497–523.

Douglas, R., and J. Partridge. "Far-Red Sensitivity in the Deep-Sea Dragon Fish *Aristostomias titmannii*." *Nature* 37 (1995): 21–22.

———. "Visual Adaptations to the Deep Sea." In *Encyclopedia of Fish Physiology: From Genome to Environment*. Orlando, FL: Academic Press, 2011.

Hecht, S., et al. "Energy, Quanta, and Vision." *Journal of General Physiology* 25 (1942): 819–40.

Herring, P. J. "Bioluminescence of Marine Organisms." *Nature* 267 (1977): 788–93.

Ings, Simon. *The Eye: A Natural History*. London: Bloomsbury, 2007.

Nagel, Thomas. "What Is It Like to Be a Bat?" *Philosophical Review* 83, no. 4 (1974): 435–50.

Nordby, Knut. "Vision in a Complete Achromat: A Personal Account." In *Night Vision: Basic, Clinical and Applied Aspects*, edited by R. F. Hess, L. T. Sharpe, and K. Nordby. Cambridge: Cambridge University Press, 1990.

O'Carroll, D. C., and E. J. Warrant. "Vision in Dim Light: Highlights and Challenges." *Philosophical Transactions of the Royal Society B* 372 (2017).

Robison, B. H., and K. R. Reisenbichler. "*Macropinna microstoma* and the Paradox of Its Tubular Eyes." *Copeia* 4 (2008): 780–84.

Sacks, Oliver. *The Island of the Color-Blind*. London: Picador, 1986.

Sharpe, L., and A. Stockman. "Rod Pathways: The Importance of Seeing Nothing." *Trends in Neuroscience* 22, no. 11 (1999): 497–504.

Stockman, A., et al. "Slow and Fast Pathways in the Human Rod Visual System: Electrophysiology and Psychophysics." *Journal of Optical Society of America* 8, no. 10 (1991): 1657–65.

Tinsley, J. N., et al. "Direct Detection of a Single Photon by Humans." *Nature Communications* 7 (2016).

Wagner, H. J., et al. "A Novel Vertebrate Eye Using Both Refractive and Reflective Optics." *Current Biology* 19 (2009): 108–14.

CHAPTER THREE
The Great Gray Owl and Our Sense of Hearing

Interviews and personal communications with Nigel Peake, Christine Köppl, and Ulrike Langemann, among others.

Beranek, Leo. "Interview of Leo Beranek by Jack Purcell on February 26, 1989." Niels Bohr Library and Archives. American Institute of Physics, 2012.

Bhatia, Aatish. "The Sound So Loud That It Circled the Earth Four Times." 2014. Nautil.us website.

Birkhead, Tim. *Bird Sense: What It's Like to Be a Bird*. London: Bloomsbury, 2012.

Cage, John. "John Cage's Lecture 'Indeterminacy,' 5'00" to 6'00."" In *Die Reihe*, edited by Herbert Eimert and Karlheinz Stockhausen. Malvern, PA: Theodore Presse, 1961.

———. *Silence: Lectures and Writing*. Middletown, CT: Wesleyan University Press, 1961.

von Campenhausen, M., and H. Wagner. "Influence of the Facial Ruff on the Sound-Receiving Characteristics of the Barn Owl's Ears." *Journal of Comparative Physiology A* 192 (2006): 1073–82.

Clark, Ian A., et al. "Bio-Inspired Canopies for the Reduction of Roughness Noise." *Journal of Sound and Vibration* 385 (2016): 33–54.

Diderot, Denis. "Letter on the Blind, 1749." In *Early Philosophical Works*, translated by M. Jourdain, 1916.

Frazer, James George. *The Fasti of Ovid*. London, 1929.

Gatty, Harold. *Finding Your Way Without Map or Compass*. Glasgow: William Collins Sons & Co., 1958.

Hausfeld, Steven, et al. "Echo Perception of Shape and Texture by Sighted Subjects." *Perceptual and Motor Skills* 55, no. 2 (1982): 623–32.

Horowitz, Seth. *The Universal Sense: How Hearing Shapes the Mind*. London: Bloomsbury, 2012.

Hull, John M. *Touching the Rock: An Experience of Blindness*. London: Society for Promoting Christian Knowledge, 1990.

Jobling, James A. *The Helm Dictionary of Scientific Bird Names: From Aalge to Zusii*. London: Christopher Helm, 2010.

Keller, Helen. "Letter to Dr. Kerr Love on 31st March 1910." In *Helen Keller in Scotland: A Personal Record Written by Herself*. London: Methuen, 1933.

Knudsen, E., and M. Konishi. "Mechanisms of Sound Localisation in the Barn Owl." *Journal of Comparative Physiology A* 133 (1979): 13–21.

Knudsen, Eric. "The Hearing of the Barn Owl." *Scientific American* 245, no. 6 (1981): 113–25.

Konishi, M. "How the Owl Tracks Its Prey: Experiments with Trained Barn Owls Reveal How Their Acute Sense of Hearing Enables Them to Catch Prey in the Dark." *American Scientist* 61, no. 4 (1973): 414–24.

Köppl, C., et al. "An Auditory Fovea in the Barn Owl Cochlea." *Journal of Comparative Physiology A* 171 (1993): 695–704.

Krumm, B., et al. "Barn Owls Have Ageless Ears." *Proceedings of the Royal Society B* 284 (2017).

Makous, J. C., and J. C. Middlebrooks. "Two-Dimensional Sound Localisation by Human Listeners." *Journal of the Acoustical Society* 87 (1990): 2188.

Martin, Graham. "Absolute Visual Threshold and Scotopic Spectral Sensitivity in the Tawny Owl *Strix aluco*." *Nature* 268 (1977): 636–38.

———. *Birds by Night*. London: Bloomsbury, 1990.

Michelson Foy, George. "I've Been to the Quietest Place on Earth." *Guardian*, May 18, 2012.

Payne, Roger. "Acoustic Location of Prey by Barn Owls, *Tyto alba*." *Journal of Experimental Biology* 54 (1971): 535–73.

Rosenblum, Lawrence D. *See What I'm Saying: The Extraordinary Powers of Our Five Senses*. New York: Norton, 2010.

Sacks, Oliver. "The 'Dark, Paradoxical Gift.'" *New York Review of Books*, April 11, 1991.

———. *Musicophilia: Tales of Music and the Brain*. London: Picador, 2007.

Smith, Dwight G. *Great Horned Owls*. Mechanicsburg, PA: Stackpole Books, 2002.

Stevens, S. S., and E. B. Newman. "The Localization of Actual Sources of Sound." *American Journal of Psychology* 48, no. 2 (1936): 297–306.

Supa, Michael, et al. "'Facial Vision': The Perception of Obstacles by the Blind." *American Journal of Psychology* 57 (1944): 133–83.

Symons, G. J., ed. "The Eruption of Krakatoa and Subsequent Phenomena." Report of the Krakatoa Committee of the Royal Society, Harrison and Sons, 1888.

Winchester, Simon. *Krakatoa: The Day the World Exploded: August 27, 1883*. New York: Harper Perennial, 2005.

References

CHAPTER FOUR
The Star-Nosed Mole and Our Sense of Touch

Interviews and personal communications with Ken Catania, Daniel Goldreich, Eşref Armağan, and Alvaro Pascual-Leone, among others.

Aristotle. *De Anima.* Translated by R. D. Hicks. Cambridge: Cambridge University Press, 1907.

Amedi, A., et al. "Neural and Behavioral Correlates of Drawing in an Early Blind Painter: A Case Study." *Brain Research* 1242 (2008): 252–62.

Blakeslee, Sandra, and Matthew Blakeslee. *The Body Has a Mind of Its Own.* New York: Penguin Random House, 2008.

Catania, K. C. "Structure and Innervation of the Sensory Organs on the Snout of the Star-Nosed Mole." *Journal of Comparative Neurology* 351 (1995): 536–48.

———. "Ultrastructure of the Eimer's Organ of the Star-Nosed Mole." *Journal of Comparative Neurology* 365 (1996): 343–54.

———. "A Nose That Looks Like a Hand and Acts Like an Eye: The Unusual Mechanosensory System of the Star-Nosed Mole." *Journal of Comparative Physiology A* 185 (1999): 367–72.

———. "Quick Guide: Star-Nosed Moles." *Current Biology* 15, no. 21 (2005): 863–64.

———. "Underwater 'Sniffing' by Semi-Aquatic Mammals." *Nature* 444 (2006): 1024–25.

Catania, K. C., and F. E. Remple. "Asymptotic Prey Profitability Drives Star-Nosed Moles to the Foraging Speed Limit." *Nature* 433 (2005): 519–22.

Catania, K. C., and J. H. Kaas. "Organization of the Somatosensory Cortex of the Star-Nosed Mole." *Journal of Comparative Neurology* 351 (1995): 549–67.

———. "Somatosensory Fovea in the Star-Nosed Mole: Behavioral Use of the Star in Relation to Innervation Patterns and Cortical Representation." *Journal of Comparative Neurology* 387 (1997): 215–33.

Catania, Kenneth. *Great Adaptations: Star-Nosed Moles, Electric Eels, and Other Tales of Evolution's Mysteries Solved.* Princeton, NJ: Princeton University Press, 2020.

DeSalle, Rob. *Our Senses: An Immersive Experience.* New Haven, CT: Yale University Press, 2018.

Eimer, T. "Die Schnautze des Maulwurfs als Tastwerkzeug." *Archiv für Mikroscopische Anatomie* 7 (1871): 181–201.

Halata, Z., et al. "Friedrich Sigmund Merkel and His 'Merkel Cell,' Morphology, Development, and Physiology: Review and New Results." *Anatomical Record Part A* 271A (2003): 225–39.

Hull, John M. *Touching the Rock: An Experience of Blindness.* London: Society for Promoting Christian Knowledge, 1990.

Keller, Helen. *The Story of My Life.* New York: Doubleday, 1905.

Linden, David. *Touch: The Science of Hand, Heart, and Mind.* New York: Viking Press, 2015.

Lovecraft, H. P. "The Call of Cthulhu." *The Call of Cthulhu and Other Weird Tales.* New York: Vintage Classics, 2011.

Marsh, Henry. *Do No Harm: Stories of Life, Death and Brain Surgery.* London: Weidenfeld & Nicolson, 2014.

Merkel, F. "Tastzellen and Tastkoerperchen bei den Hausthieren und beim Menschen." *Archiv für Mikroscopische Anatomie* 11 (1875): 636–52.

Pascual-Leone, A., and F. Torres. "Plasticity of the Sensorimotor Cortex Representation of the Reading Finger In Braille Readers." *Brain* 116, no. 1 (1993): 39–52.

Pascual-Leone, A., and R. Hamilton. "The Metamodal Organization of the Brain." *Progress in Brain Research* 134 (2001): 427–45.

Penfield, W., and T. Rasmussen. *The Cerebral Cortex of Man.* New York: Macmillan, 1950.

Peters, R. M., et al. "Diminutive Digits Discern Delicate Details: Fingertip Size and the Sex Difference in Tactile Spatial Acuity." *Journal of Neuroscience* 29, no. 50 (2009): 15757–61.

"Speaking of Pictures . . . Japanese Skin Specialist Collects Human Tattoos for Tokyo Museum." *Life,* July 3, 1950.

CHAPTER FIVE

The Common Vampire Bat and Our Senses of Pleasure and Pain

Interviews and personal communications with Gerry Wilkinson, Francis McGlone, Steve Waxman, Pam Costa, Uwe and Christel Schmidt, Ludwig Kürten, and David Julius, among others.

Ackerman, Diane. *A Natural History of the Senses.* New York: Vintage Books, 1990.

Bales, K. L., et al. "Social Touch during Development: Long-Term Effects on Brain and Behavior." *Neuroscience and Biobehavioral Reviews* 95 (2018): 202–19.

Bohlen, C. J., and D. Julius. "Receptor-Targeting Mechanisms of Pain-Causing Toxins: How Ow?" *Toxicon* 60, no. 3 (2012): 254–64.

Botvinick, M., and J. Cohen. "Rubber Hands That 'Feel' Touch That Eyes See." *Nature* 391 (1998): 756.

Burbank, M. K., et al. "Familial Erythromelalgia: Genetic and Physiologic Observations." *Journal of Laboratory and Clinical Medicine* 68, no. 5 (1966): 861.

Cox, J. J., et al. "An SCN9A Channelopathy Causes Congenital Inability to Experience Pain." *Nature* 444 (2006): 896–98.

Crucianelli, L., et al. "Bodily Pleasure Matters: Velocity of Touch Modulates Body Ownership During the Rubber Hand Illusion." *Frontiers of Psychology,* October 8, 2013.

Crusco, A. H., and C. G. Wetzel. "The Midas Touch: The Effects of Interpersonal Touch on Restaurant Tipping." *Personality and Social Psychology Bulletin* 10 (1984): 512–17.

References

Cummins, T. R., et al. "Electrophysiological Properties of Mutant NaV1.7 Sodium Channels in a Painful Inherited Neuropathy." *Journal of Neuroscience* 24, no. 38 (2004): 8232–36.

Dawkins, Richard. *The Selfish Gene*. 2nd ed. Oxford: Oxford University Press, 1990.

Denworth, Lydia. "The Social Power of Touch." *Scientific American Mind* 26, no. 4 (July/August 2015): 30–9.

Essick, G. E., et al. "Quantitative Assessment of Pleasant Touch." *Neuroscience and Biobehavioral Reviews* 34 (2010): 192–203.

Gallace, A., and C. Spence. "The Science of Interpersonal Touch: An Overview." *Neuroscience and Biobehavioral Reviews* 35 (2010): 246–59.

Gracheva, E. O., et al. "Molecular Basis of Infrared Detection by Snakes." *Nature* 464 (2011): 1006–11.

———. "Ganglion-Specific Splicing of TRPV1 Underlies Infrared Sensation in Vampire Bats." *Nature* 476, no. 7358 (2011): 88–91.

Gueguen, N., and Celine Jacob. "The Effect of Touch on Tipping: An Evaluation in a French Bar." *Hospitality Management* 24 (2005): 295–99.

Kürten, L., and U. Schmidt. "Thermoperception in the Common Vampire Bat (*Desmodus rotundus*)." *Journal of Comparative Physiology* 146 (1982): 223–28.

Kürten, L., et al. "Warm and Cold Receptors in the Nose of the Vampire Bat *Desmodus rotundus*." *Naturwissenschaften* 71 (1984): 327–28.

Linden, David. *Touch: The Science of Hand, Heart, and Mind*. New York: Viking Press, 2015.

Lowa, D. H. W., et al. "Dracula's Children: Molecular Evolution of Vampire Bat Venom." *Journal of Proteomics* 89 (2013): 95–111.

Lumpkin, E. A., et al. "The Cell Biology of Touch." *Journal of Cell Biology* 191, no. 2 (2010): 237–48.

Maitre, N. L., et al. "The Dual Nature of Early-Life Experience on Somatosensory Processing in the Human Infant Brain." *Current Biology* 27, no. 7 (2017): 1048–54.

McGlone, F., et al. "Discriminative and Affective Touch: Sensing and Feeling." *Neuron* 82 (2014): 737–55.

Nabokov, Vladimir. *Lolita*. New York: Olympia Press, 1955.

Newman, E. A., and P. A. Hartline. "The Infrared 'Vision' of Snakes." *Scientific American* 246 (1982): 116–27.

Perl, Edward R. "Ideas About Pain: A Historical Review." *Neuroscience* 8 (2007): 71–80.

Sachs, Frederick. "The Intimate Sense: Understanding the Mechanics of Touch." *Sciences* 28, no. 1 (January/February 1988): 28–34.

Sherrington, Charles. *The Integrative Action of the Nervous System*. New York: Scribner, 1906.

Vallbo, A., et al. "Unmyelinated Afferents Constitute a Second System Coding Tactile Stimuli of the Human Hairy Skin." *Journal of Neurophysiology* 81 (1999): 2753–63.

Vay, L., et al. "The Thermo-TRP Ion Channel Family: Properties and Therapeutic Implications." *British Journal of Pharmacology* 165, no. 4 (2012): 787–801.

Voos, A. C., et al. "Autistic Traits Are Associated with Diminished Neural Response to Affective Touch." *Social Cognitive and Affective Neuroscience* 8, no. 4 (2013): 378–86.

Waxman, S. G. "A Channel Sets the Gain on Pain." *Nature* 444 (2006): 831–32.

Waxman, Stephen. *Chasing Men on Fire: The Story of the Search for a Pain Gene.* Cambridge, MA: MIT Press, 2018.

Wilkinson, Gerald S. "Reciprocal Food Sharing in the Vampire Bat." *Nature* 308 (1984): 181–84.

———. "Social Grooming in the Common Vampire Bat, *Desmondus rotundus.*" *Animal Behaviour* 34 (1986): 1880–89.

———. "Food Sharing in Vampire Bats." *Scientific American* 262, no. 2 (1990): 76–83.

Yang, Y., et al. "Mutations in SCN9A, Encoding a Sodium Channel Alpha Subunit, in Patients with Primary Erythermalgia." *Journal of Medical Genetics* 41 (2004): 171–74.

CHAPTER SIX

The Goliath Catfish and Our Sense of Taste

Interviews and personal communications with Jelle Atema, Linda Bartoshuk, Tom Finger, and John Caprio, among others.

Ackerman, Diane. *A Natural History of the Senses.* New York: Vintage Books, 1990.

Atema, Jelle. "Structures and Functions of the Sense of Taste in the Catfish *Ictalurus natalis.*" *Brain, Behavior and Evolution* 4 (1971): 273–94.

———. "Smelling and Tasting Underwater." *Oceanus* 23, no. 3 (1980): 4–18.

Bartoshuk, L. "Taste." In *Stevens' Handbook of Experimental Psychology,* edited by R. C. Atkinson et al. New York: Wiley, 1988.

———. "Ratio-Scaling, Taste Genetics and Taste Pathologies." In *Ratio Scaling of Psychological Magnitude,* edited by S. J. Bolanowski and G. A. Gescheider. Mahwah, NJ: Erlbaum, 1991.

———. "Sweetness: History, Preference and Genetic Variability." *Food Technology* 45, no. 11 (1991): 108–13.

Bartoshuk, L., et al. "What Aristotle Didn't Know About Flavor." *American Psychologist* 74, no. 9 (2019): 1003–11.

Bartoshuk, L. M., et al. "PTC/PROP Tasting: Anatomy, Psychophysics, and Sex Effects." *Physiology and Behavior* 56 (1994): 1165–71.

Basson, M. D., et al. "Association between 6-n-propylthiouracil (PROP) Bitterness and Colonic Neoplasms." *Digestive Diseases and Sciences* 50, no. 3 (2005): 483–89.

Brillat-Savarin, Jean Anthelme. *Physiology of Taste; Or, Meditations on Transcendental Gastronomy: Theoretical, Historical and Practical Work* (1825). Merchant Books, 2009.

Caprio, John. "Electrophysiological Distinctions Between Taste and Smell of Amino Acids in Catfish." *Nature* 266 (1977): 850–51.

———. "Marine Teleost Locates Live Prey Through pH Sensing." *Science* 344, no. 6188 (2014): 1154–56.

Chaudari, N., et al. "The Taste of Monosodium Glutamate: Membrane Receptors in Taste Buds." *Journal of Neuroscience* 16, no. 12 (1996): 3817–26.

Citron, F. M. M., and A. E. Goldberg. "Metaphorical Sentences Are More Engaging Than Their Literal Counterparts." *Journal of Cognitive Neuroscience* 26, no. 11 (2014): 2585–95.

DeSalle, Rob. *Our Senses: An Immersive Experience.* New Haven: Yale University Press, 2018.

Essick, G. E., et al. "Lingual Tactile Acuity, Taste Perception, and the Density and Diameter of Fungiform Papillae in Female Subjects." *Physiology and Behavior* 80 (2003): 289–302.

Finger, Thomas, and Sue Kinnamon. "Taste Isn't Just for Taste Buds Anymore." *F1000 Biology Report* 3 (2011).

Finger, Thomas E. "Gustatory Pathways in the Bullhead Catfish. I. Connections of the Anterior Ganglion." *Journal of Comparative Neurology* 165 (1976): 513–26.

———. "Gustatory Pathways in the Bullhead Catfish. II. Facial Lobe Connections." *Journal of Comparative Neurology* 180 (1978): 691–705.

———. "Solitary Chemoreceptor Cells in the Nasal Cavity Serve as Sentinels of Respiration." *Proceedings of the National Academy of Sciences* 100 (2003): 8981–86.

———. "Evolution of Taste from Single Cells to Taste Buds." *Chemosense* 9, no. 33 (2009): 1–6.

Gilbert, Avery. *What the Nose Knows: The Science of Scent in Everyday Life.* Synesthetics, 2014.

Goode, Erica. "If Things Taste Bad, 'Phantoms' May Be at Work." *New York Times*, April 13, 1999.

Herrick, C. J. "The Organ and Sense of Taste in Fishes." *Bulletin of US Fishery Committee* 22 (1904): 237–72.

Höfer, D., et al. "Taste Receptor-Like Cells in the Rat Gut Identified by Expression of Alpha-Gustducin." *Proceedings of the National Academy of Sciences* 25 (1996): 6631–34.

Ikeda, Kikunae. "New Seasonings." *Journal of Tokyo Chemical Society* 30 (1909): 820–36.

Malik, B., et al. "Mammalian Taste Cells Express Functional Olfactory Receptors." *Chemical Senses* 20 (2019): 1–13.

Roosevelt, Theodore. *Through the Brazilian Wilderness.* New York: Charles Scribner's Sons, 1914.

Running, C., et al. "Oleogustus: The Unique Taste of Fat." *Chemical Senses* 40, no. 7 (2015): 507–16.

Shepherd, Gordon M. *Neurogastronomy: How the Brain Creates Flavor and Why It Matters.* New York: Columbia University Press, 2012.

Steiner, Jacob. "Facial Expressions of the Neonate Infant Indicating the Hedonics of Food-Related Chemical Stimuli." In *Taste and Development: The Genesis of Sweet Preference*, edited by James Weiffenbach. Washington, DC: National Institutes of Health, 1977.

Whitear, M. "Solitary Chemoreceptor Cells." In *Chemoreception in Fishes*, edited by T. J. Hara. 2nd ed. Amsterdam: Elsevier, 1992.

Yanagisawa, K., et al. "Anesthesia of the Chorda Tympani Nerve and Taste Phantoms." *Physiology and Behavior* 63, no. 3 (1998): 329–35.

CHAPTER SEVEN
The Bloodhound and Our Sense of Smell

Interviews and personal communications with Gary Settles, Matthias Laska, Andreas Keller, and John McGann, among others.

Brey, Catherine, and Lena Reed. *The New Complete Bloodhound.* Nashville, TN: Howell Book House, 1978.

Broca, Pau. "Recherches sur les centres olfactifs." *Revue d'Anthropologie* 2 (1879): 390–91.

Buck, L., and R. Axel. "A Novel Multigene Family May Encode Odorant Receptors: A Molecular Basis for Odor Recognition." *Cell* 65 (1991): 175–87.

Bushdid, C., et al. "Humans Can Discriminate More Than 1 Trillion Olfactory Stimuli." *Science* 343, no. 6177 (2014): 1370–72.

Chesterton, G. K., *Wine, Water, and Song.* London: Methuen, 1915.

Coren, Stanley. *How Dogs Think: Understanding the Canine Mind.* New York: Simon & Schuster, 2008.

Craven, B. A., et al. "The Fluid Dynamics of Canine Olfaction: Unique Nasal Airflow Patterns as an Explanation of Macrosmia." *Journal of the Royal Society Interface* 7 (2010): 933–43.

Curran, A. M., et al. "Canine Human Scent Identifications with Post-Blast Debris Collected from Improvised Explosive Devices." *Forensic Science International*, 199 (2010): 103–108.

Feynman, Richard. *"Surely You're Joking, Mr. Feynman!": Adventures of a Curious Character.* New York: Norton, 1985.

Firestein, Stuart. "How the Olfactory System Makes Sense of Scents." *Nature* 413 (2001): 211–17.

Freud, Sigmund. *Minutes de la Société Psychanalytique de Vienne, November 17, 1909.* Paris: Gallimard, 1978.

Gilbert, Avery. *What the Nose Knows: The Science of Scent in Everyday Life.* Synesthetics, 2014.

Glausiusz, Josie. "Raw Data: Scents and Scents-Ability." *Discover Magazine*, March 15, 2007.

Herz, Rachel. *The Scent of Desire: Discovering our Enigmatic Sense of Smell*. New York: Morrow, 2007.

Horowitz, Alexandra. *Being a Dog: Following the Dog into a World of Smell*. New York: Simon & Schuster, 2016.

Keller, Helen. *The World I Live In*. New York: Century Co., 1908.

Laing, D. G. "Natural Sniffing Gives Optimum Odor Perception for Humans." *Perception* 12 (1983): 99–117.

Laska, M. "Busting a Myth: Humans Are Not Generally Less Sensitive to Odors Than Nonhuman Mammals." *Chemical Senses* 40 (2015): 537.

———. "Human and Animal Olfactory Capabilities Compared." In *The Springer Handbook of Odor*, edited by A. Buettneer. New York: Springer, 2017.

Lawson, M. J., et al. "A Computational Study of Odorant Transport and Deposition in the Canine Nasal Cavity: Implications for Olfaction." *Chemical Senses* 37 (2012): 553–66.

Mainland, Joel, and Noam Sobel. "The Sniff Is Part of the Olfactory Percept." *Chemical Senses* 31, no. 2 (2006): 181–96.

McGann, John P. "Poor Human Olfaction Is a 19th-Century Myth." *Science* 356, no. 6338 (2017): 7263–69.

Oliveira-Pinto, A. V., et al. "Sexual Dimorphism in the Human Olfactory Bulb: Females Have More Neurons and Glial Cells Than Males." *PLoS One* 9 (2014).

Porter, J., et al. "Mechanisms of Scent-Tracking in Humans." *Nature Neuroscience* 10 (2007): 27–29.

Ribeiro, P. F., et al. "Greater Addition of Neurons to the Olfactory Bulb Than to the Cerebral Cortex of Eulipotyphlans But Not Rodents, Afrotherians or Primates." *Frontiers in Neuroanatomy* 8 (2014): 23, 50.

Sacks, Oliver. "Interview with Adam Higginbotham for *Telegraph Magazine*," November 2012.

———. "The Dog Beneath the Skin." In *The Man Who Mistook His Wife for a Hat*. London: Picador, 1986.

Sagan, Carl. *Broca's Brain: Reflections on the Romance of Science*. New York: Random House, 1979.

Settles, Gary. "Sniffers: Fluid Dynamic Sampling for Olfactory Trace Detection in Nature and Homeland Security." *Journal of Fluids Engineering* 127 (2005): 189–218.

Settles, G. S., et al. "The External Aerodynamics of Canine Olfaction." In *Sensors and Sensing in Biology and Engineering*, edited by F. G. Barth et al. Vienna: Springer, 1996.

Shepherd, Gordon M. "The Human Sense of Smell: Are We Better Than We Think?" *PLoS Biology* 2, no. 5 (2004).

———. *Neurogastronomy: How the Brain Creates Flavor and Why It Matters*. New York: Columbia University Press, 2012.

Walker, James, et al. "Human Odor Detectability: New Methodology Used to Determine Threshold and Variation." *Chemical Senses* 28 (2003): 817–26.

———. "Naturalistic Quantification of Canine Olfactory Sensitivity." *Applied Animal Behaviour Science* 97, no. 2–4 (May 2006): 241–54.

Whitney, Leon F. *Bloodhounds and How to Train Them*. New York: Orange Judd Publishing, 1947.

CHAPTER EIGHT
The Giant Peacock of the Night and Our Sense of Desire

Interviews and personal communications with Karl-Ernst Kaissling, Tristram Wyatt, David Booth, Tamsin Saxton, and Claus Wedekind, among others.

Butenandt, Adolf, et al. "Uber den Sexual-Lockstoff des Seidenspinners Bombyx mori. Reindarstellung und Konstitution." *Z. Naturforsch* 14 (1959): 283–84.

Cowley, J. J., and B. W. L. Brooksbank. "Human Exposure to Putative Pheromones and Changes in Aspects of Social Behavior." *Journal of Steroid Biochemistry and Molecular Biology* 39, no. 4/2 (1991): 647–59.

Dawkins, Richard. *The Selfish Gene*. New York: Oxford University Press, 1976.

Doty, Richard. *The Great Pheromone Myth*. Baltimore, MD: Johns Hopkins University Press, 2010.

Doucet, S., et al. "The Secretion of Areolar (Montgomery's) Glands from Lactating Women Elicits Selective, Unconditional Responses in Neonates." *PLoS One*, October 23, 2009.

Ellis, Havelock. *Studies on the Psychology of Sex* Vol. 4, *Sexual Selection in Man*. New York: Random House, 1927.

Fabre, Jean-Henri. *The Life of the Caterpillar*. Translated by Alexander Teixeira de Mattos. New York: Dodd, Mead, 1916.

Forbes, William T. M. "The Silkworm of Aristotle." *Classical Philology* 25, no. 1 (1930): 22–26.

"Henri Fabre Dies in France at 92." *New York Times*, October 11, 1915.

Herz, Rachel. *The Scent of Desire: Discovering Our Enigmatic Sense of Smell*. New York: Morrow, 2007.

———. "The Truth about Pheromones, Part 2." *Psychology Today*, June 18, 2009.

Horn, Miriam. *Rebels in White Gloves: Coming of Age with the Wellesley Class of '69*. New York: Crown, 1999.

Kaissling, Karl-Ernst. "Pheromone-Controlled Anemotaxis in Moths." In *Orientation and Communication in Arthropods*, edited by M. Lehrer. Basel: Birkhauser Verlag, 1997.

———. "Pheromone Reception in Insects." In *Neurobiology of Chemical Communication*, edited by D. Mucignat-Caretta. Boca Raton, FL: CRC Press, 2014.

Karlson, Peter, and Adolf Butenandt. "Pheromones (Ectohormones) in Insects." *Annual Review of Entomology* 4 (1959): 39.

Karlson, Peter. "Obituary of Adolf Butenandt." *Independent*, February 1, 1995.

References

Karlson, Peter, and Martin Lüscher. "'Pheromones': A New Term for a Class of Biologically Active Substances." *Nature* 183 (1959): 55–56.

Kirk-Smith, M. D., and D. A. Booth. "Effect of Androstenone on Choice of Location in Others' Presence." In *Olfaction and Taste VII*, edited by H. van der Starre. London: Information Retrieval, 1980.

McClintock, M. K. "Menstrual Synchrony and Suppression." *Nature* 229 (1971): 244–45.

Miller, G., et al. "Ovulatory Cycle Effects on Tip Earnings by Lap Dancers: Economic Evidence of Human Estrus?" *Evolution and Human Behavior* 28 (2007): 375–81.

Mujica-Parodi, L., et al. "Second-Hand Stress: Neurobiological Evidence for a Human Alarm Pheromone." *Nature Precedings*, November 25, 2008.

Müller-Schwarzem, Dietland, et al. "Alert Odor from Skin Gland in Deer." *Journal of Chemical Ecology* 10, no. 12 (1984): 1707–29.

Ober, Carole, et al. "HLA and Mate Choice in Humans." *American Journal of Human Genetics* 61 (1997): 497–504.

Pause, B. "Are Androgen Steroids Acting as Pheromones in Humans?" *Physiology and Behavior* 83 (2004): 21–29.

Preti, George. "Human Pheromones: What's Purported, What's Supported." Report for Sense of Smell Institute, July 2009.

Rau, P., and N. L. Rau. "The Sex Attraction and Rhythmic Periodicity in Giant Saturniid Moths." *Transactions of the Academy of Sciences of St. Louis* 26 (1929): 83–221.

Rodriguez, I., et al. "A Putative Pheromone Receptor Gene Expressed in Human Olfactory Mucosa." *Nature Genetics* 26 (2000): 18–19.

Rosenblum, Lawrence D. *See What I'm Saying: The Extraordinary Powers of Our Five Senses.* New York: Norton, 2011.

Saxton, Tamsin, et al. "Evidence That Androstadienone, a Putative Human Chemosignal, Modulates Women's Attributions of Men's Attractiveness." *Hormones and Behavior* 54 (2008): 597–601.

Stern, K., and M. K. McClintock. "Regulation of Ovulation by Human Pheromones." *Nature* 392 (1998): 177–79.

Stoddart, D. Michael. *The Scented Ape: The Biology and Culture of Human Odor.* Cambridge: Cambridge University Press, 1990.

Süskind, Patrick. *Perfume: The Story of a Murderer.* London: Hamish Hamilton, 1986.

Thomas, Lewis. *The Lives of a Cell: Notes of a Biology Watcher.* New York: Viking Press, 1974.

Wedekind, Claus, and S. Füri. "Body Odor Preferences in Men and Women: Do They Aim for Specific MHC Combinations or Simply Heterozygosity?" *Proceedings of the Royal Society of London B* 264 (1997): 1471–79.

Wedekind, Claus, et al. "MHC-Dependent Mate Preferences in Humans." *Proceedings of the Royal Society of London B* 260 (1995): 245–49.

"What Don't We Know." *Science* 309 (2005): 93.

"Who Made Speed Dating?" *New York Times*, September 29, 2013.

Witt, M., and T. Hummel. "Vomeronasal versus Olfactory Epithelium: Is There a Cellular Basis for Human Vomeronasal Perception?" *International Review of Cytology* 248 (2006): 209–59.

Wyatt, Tristram D. "Fifty Years of Pheromones." *Nature* 457 (2009): 262–63.

———. *Pheromones and Animal Behavior: Chemical Signals and Signatures.* Cambridge: Cambridge University Press, 2014.

———. "Primer: Pheromones." *Current Biology* 27 (2017): 523–67.

Wysocki, C. J., and G. Preti. "Pheromones in Mammals." In *Encyclopedia of Neuroscience*, edited by Larry Squire. Orlando, FL: Academic Press, 2009.

CHAPTER NINE
The Cheetah and Our Sense of Balance

Interviews and personal communications with Alan Wilson, Brian Day, Camille Grohé, Fred Spoor, and Barry Seemungal, among others.

Berthoz, Alain. *The Brain's Sense of Movement.* Cambridge, MA: Harvard University Press, 2000.

Cox, P. G., and N. Jeffrey. "Semicircular Canals and Agility: The Influence of Size and Shape Measures." *Journal of Anatomy* 216 (2010): 37–47.

Day, Brian, and Richard Fitzpatrick. "Primer: The Vestibular System." *Current Biology* 15, no. 15 (2005): R583–86.

Douglas, Ron. "Acoustic Neuroma and Its Ocular Implications: A Personal View." *Optometry*, January 25, 2002.

Fitzpatrick, R. C., and B. L. Day. "Probing the Human Vestibular System with Galvanic Stimulation." *Journal of Applied Physiology* 96 (2004): 2301–16.

Fitzpatrick, R. C., et al. "Resolving Head Rotation for Bipedalism." *Current Biology* 16 (2006): 1509–14.

Grohé, C., et al. "Recent Inner Ear Specialization for High-Speed Hunting in Cheetahs." *Nature Scientific Reports*, February 2, 2018.

Hadžiselimović, H., and L. J. Savković. "Appearance of Semicircular Canals in Birds in Relation to Mode of Life." *Acta Anatomica* 57 (1964): 306–15.

Hudson, P. E., et al. "High Speed Galloping in the Cheetah (*Acinonyx jubatus*) and the Racing Greyhound (*Canis familiaris*): Spatio-Temporal and Kinetic Characteristics." *Journal of Experimental Biology* 215 (2012): 2425–34.

Latimer, B., and C. Owen Lovejoy. "The Calcaneus of *Australopithecus afarensis* and Its Implications for the Evolution of Bipedality." *American Journal of Physical Anthropology* 78, no. 3 (1989): 369–86.

Nigmatullina, Y., et al. "The Neuroanatomical Correlates of Training-Related Perceptuo-Reflex Uncoupling in Dancers." *Cerebral Cortex* 25 (2015): 554–62.

Phillips, Helen. "The Cheetah's Time Has Come." *Nature* 386 (1997): 653.

Piccolino, Marco. "The Bicentennial of the Voltaic Battery (1800–2000): The Artificial Electric Organ." *Trends in Neurosciences* 23, no. 4 (2000): 147–51.

Sacks, Oliver. "On the Level." In *The Man Who Mistook His Wife for a Hat*. London: Picador, 1986.

Sharp, N. C. C. "Timed Running Speed of a Cheetah (*Acinonyx jubatus*)." *Journal of Zoology* 241 (1997): 493–94.

Smith, Roff. "Cheetah Breaks Speed Record—Beats Usain Bolt by Seconds." *National Geographic News*, August 2, 2012.

Spoor, F., et al. "Implications of Early Hominid Labyrinthine Morphology for Evolution of Human Bipedal Locomotion." *Nature* 369 (1994): 645–48.

———. "The Primate Semicircular Canal System and Locomotion." *Proceedings of the National Academy of Sciences* 104, no. 26 (2007): 10808–12.

Tuttle, R. H. "Kinesiological Inferences and Evolutionary Implications from Laetoli Bipedal Trails G-1, G-2/3, and A." In *Laetoli: A Pliocene Site in Northern Tanzania*, edited by M. D. Leakey and J. M. Harris. London: Clarendon Press, 1987.

Williams T. M., et al. "Skeletal Muscle Histology and Biochemistry of an Elite Sprinter, the African Cheetah." *Journal of Comparative Physiology B* 167 (1997): 527–35.

Wilson A., et al. "Locomotion Dynamics of Hunting in Wild Cheetahs." *Nature* 498 (2013): 185–88.

CHAPTER TEN
The Trashline Orbweaver and Our Sense of Time

Interviews and personal communications with Thomas Jones, Darrell Moore, Mark Threadgold, Russell Foster, and Ron Douglas (again), among others.

Arnold, Carrie. "Spiders Listen to Their Webs." *National Geographic*, June 5, 2014.

Aschoff, Jürgen. "Circadian Rhythms in Man." *Science* 148 (1965): 1427–32.

Boyle, Rebecca. "Smallest Sliver of Time Yet Measured Sees Electrons Fleeing Atoms." *New Scientist*, November 19, 2016.

Daley, Jason. "Meet the Zeptosecond, the Smallest Slice of Time Yet Recorded." *Smithsonian Magazine*, November 15, 2016.

Douglas, Ron, and Russell Foster. "The Eye: Organ of Space and Time." *Optician*, March 20, 2015.

Foster, R. G., and L. Kreitzman. *Rhythms of Life*. New Haven, CT: Yale University Press, 2004.

———. "The Rhythms of Life: What Your Body Clock Means to You!" *Experimental Physiology* 99, no. 4 (2014): 599–606.

———. *Circadian Rhythms: A Very Short Introduction*. Oxford: Oxford University Press, 2017.

Foster, R. G., et al. "Circadian Photoreception in the Retinally Degenerate Mouse (rd/rd)." *Journal of Comparative Physiology A* 169 (1991): 39–50.

Globig, Michel. "A World Without Day or Night." *Max Planck Research* 2 (2007).

Hattar, S., et al. "Melanopsin-Containing Retinal Ganglion Cells: Architecture, Projections, and Intrinsic Photosensitivity." *Science* 295 (2002): 1065–70.

———. "Melanopsin and Rod-Cone Photoreceptive Systems Account for All the Major Accessory Visual Functions in Mice." *Nature* 424 (2003): 76–81.

"J. J. O. de Mairan Eulogy." *Histoire de l'Académie Royale des Sciences* (1771): 89–104.

Klarsfeld, André. "At the Dawn of Chronobiology." Translated by Helen Tomlinson. September 2013.

Kleitman, Nathaniel. *Sleep and Wakefulness*. Chicago: University of Chicago Press, 1963.

Lucas, Robert J., et al. "Regulation of the Mammalian Pineal by Non-Rod, Non-Cone, Ocular Photoreceptors." *Science* 284 (1999): 505–507.

de Mairan, J. J. O. "Observation botanique." *Histoire de l'Académie Royale des Sciences* (1729): 35–36.

Moore, Darrell, et al. "Exceptionally Short-Period Circadian Clock in *Cyclosa turbinata*: Regulation of Locomotor and Web-Building Behavior in an Orb-Weaving Spider." *Journal of Arachnology* 44 (2016): 388–96.

Mortimer, Beth, et al. "The Speed of Sound in Silk, Linking Material Performance to Biological Function." *Advanced Materials* 26, no. 30 (2014): 5179–83.

du Noüy, L. C. *Biological Time*. London: Macmillan, 1937.

Ossiander, M., et al. "Attosecond Correlation Dynamics." *Nature Physics* 13 (2017): 280–85.

Pittendrigh, Colin S. "VIII. Biological Clocks: The Functions of Ancient and Modern, of Circadian Oscillations." In *Science in the Sixties: The Tenth Anniversary AFOSR Scientific Seminar, June 1965*, edited by David L. Arm. Albuquerque: University of New Mexico, 1965.

Roenneberg, T., and M. Merrow. "Circadian Clocks: The Fall and Rise of Physiology." *Nature Molecular Cell Biology* 6 (2005): 965–71.

Roenneberg, Till. *Internal Time: Chronotypes, Social Jet Lag and Why You're So Tired*. Cambridge, MA: Harvard University Press, 2017.

Siffre, Michel. *Beyond Time*. New York: McGraw-Hill, 1964.

———. "Caveman: An Interview with Michel Siffre." *Cabinet* 30 (Summer 2008).

Wilson, E. O. *Consilience*. New York: Knopf, 1998.

CHAPTER ELEVEN
The Bar-Tailed Godwit and Our Sense of Direction

Interviews and personal communications with Bob Gill and Lee Tibbitts, Henrik Mouritsen, and Joe Kirschvink, among others.

Baker, Robin. "Goal Orientation by Blindfolded Humans After Long-Distance Displacement: Possible Involvement of a Magnetic Sense." *Science* 210, no. 4469 (1980): 555–57.

————. *Human Navigation and the Sixth Sense.* New York: Touchstone, 1981.

————. *Human Navigation and Magnetoreception.* 30th anniv. ed. Manchester, UK: Manchester University Press, 1989.

Begall, Sabine, et al. "Magnetic Alignment in Grazing and Resting Cattle and Deer." *Proceedings of the National Academy of Sciences* 105, no. 36 (2008): 13451–55.

Birkhead, Tim. *Bird Sense: What It's Like to Be a Bird.* London: Bloomsbury, 2012.

Boroditsky, Lera. "How Language Shapes the Way We Think." TED video, November 2017.

Brown, F. A., et al. "A Magnetic Compass Response of an Organism." *Biological Bulletin* 119, no. 1 (1960): 367–81.

Cresci, Alessandro, et al. "Glass Eels (*Anguilla anguilla*) Have a Magnetic Compass Linked to the Tidal Cycle." *Science Advances* 3, no. 9 (2017): 1–8.

Darwin, C. R. "Origin of Certain Instincts." *Nature: A Weekly Illustrated Journal of Science* 7 (1873): 417–18.

Deutscher, Guy. *Through the Language Glass: Why the World Looks Different in Other Languages.* London: Arrow, 2011.

Driscoll, P. V., and M. Ueta. "The Migration Route and Behavior of Eastern Curlews *Numenius madagascariensis.*" *Ibis* 144 (2002): E119–30.

Egevang, E., et al. "Tracking of Arctic Terns *Sterna paradisaea* Reveals Longest Animal Migration." *Proceedings of the National Academy of Sciences* 107, no. 5 (2010): 2078–81.

Emlen, S. T. "Migratory Orientation in the Indigo Bunting, *Passerina cyanea*, Part I: Evidence for Use of Celestial Cues." *Auk* 84, no. 3 (1967): 309–42.

————. "Migratory Orientation in the Indigo Bunting, *Passerina cyanea*, Part II: Mechanism for Celestial Orientation." *Auk* 84, no. 4 (1967): 463–89.

Engels, S., et al. "Anthropogenic Electromagnetic Noise Disrupts Magnetic Compass Orientation in a Migratory Bird." *Nature* 509 (2014): 353–56.

Foley, Lauren E., et al. "Human Cryptochrome Exhibits Light-Dependent Magnetosensitivity." *Nature Communications* 2, no. 1 (2011): 1–3.

Gill Jr., Robert E., et al. "Extreme Endurance Flights by Landbirds Crossing the Pacific Ocean: Ecological Corridor Rather Than Barrier?" *Proceedings of the Royal Society* 276 (2009): 447–57.

Gould, J. L., and K. P. Able. "Human Homing: An Elusive Phenomenon." *Science* 212 (1981): 1061.

Guerra, P. A., et al. "A Magnetic Compass Aids Monarch Butterfly Migration." *Nature Communications* 5, no. 4164 (2014): 1–8.

Hart, Vlastimil, et al. "Dogs Are Sensitive to Small Variations of the Earth's Magnetic Field." *Frontiers in Zoology* 10, no. 80 (2013).

Hiscock, Hamish G., et al. "The Quantum Needle of the Avian Magnetic Compass." *Proceedings of the National Academy of Sciences* 113, no. 17 (2016): 4634–39.

Kimchi, Tali, and Joseph Terkel. "Magnetic Compass Orientation in the Blind Mole Rat." *Journal of Experimental Biology* 204 (2001): 751–78.

References

Kirschvink, J. L., et al. "Magnetite Biomineralization in the Human Brain." *Proceedings of the National Academy of Science* 89, no. 16 (1992): 7683–87.

Kirschvink, Joe. "Radio Waves Zap the Biomagnetic Compass." *Nature* 509 (2014): 296–97.

Kramer, Gustav. "Experiments on Bird Orientation." *Ibis* 94, no. 2 (1952): 265–85.

Lohmann, Kenneth J., et al. "Geomagnetic Map Used in Sea-Turtle Navigation." *Nature* 428 (2004): 909–10.

Lohmann, Kenneth J. "Magnetic Remanence in the Western Atlantic Spiny Lobster." *Journal of Experimental Biology* 113 (1984): 29–41.

Merkel, F. W., and W. Wiltschko. "Magnetismus und Richtungsfinden sugunruhiger Rotkehlchen." *Vogelwarte* 23 (1965): 71–77.

Mouritsen, Henrik. "Long-Distance Navigation and Magnetoreception in Migratory Animals." *Nature* 558 (2018): 50–59.

Phillips, John. "Magnetic Compass Orientation in the Eastern Red-Spotted Newt." *Journal of Comparative Physiology A* 158 (1986): 103–109.

Putnam, Nathan F., et al. "Evidence for Geomagnetic Imprinting as Homing Mechanism in Pacific Salmon." *Current Biology* 23 (2013): 312–16.

Ritz, T., and K. Schulten. "A Model for Photoreceptor-Based Magnetoreception in Birds." *Journal of Biophysics* 78, no. 2 (2000): 707–18.

Schulten, K. "Our Brains Might Sense Earth's Magnetic Field Like Birds." *New Scientist* 18 (2019).

Servick, K. "Humans May Sense Earth's Magnetic Field." *Science* 363, no. 6433 (2019): 1257–58.

Solov'yov, I. A., et al. "Acuity of Cryptochrome and Vision-Based Magnetoreception System in Birds." *Biophysical Journal* 99 (2010): 40–49.

Vali, H., and J. Kirschvink. "Observations of Magnetosome Organization, Surface Structure, and Iron Biomineralization of Undescribed Magnetic Bacteria: Evolutionary Speculations." *Iron Biominerals,* edited by R. B. Frankel and R. P. Blakemore. New York: Springer, 1991.

Wang, C. X., et al. "Transduction of the Geomagnetic Field as Evidenced from Alpha-Band Activity in the Human Brain." *eNeuro* 6, no. 2 (2019).

Warrant, E., et al. "The Australian Bogong Moth: A Long Distance Nocturnal Navigator." *Frontiers of Behavioral Neuroscience* 10, no. 77 (2016): 1–17.

Westby, G. W., and K. J. Partridge. "Human Homing: Still No Evidence Despite Geomagnetic Controls." *Journal of Experimental Biology* 120 (1986): 325–31.

Woodley, Keith. *Godwits: Long Haul Champions.* New Zealand: Raupo Publishing, 2009.

CHAPTER TWELVE
The Common Octopus and Our Sense of Body

Interviews and personal communications with Benny Hochner, Ian Waterman, and Jonathan Cole, among others.

References

Altman, J. S. "Control of Accept and Reject Reflexes in the Octopus." *Nature* 229 (1971): 204–206.

Aristotle. *The History of Animals, Book IX*. Translated by D'Arcy Wentworth Thompson. Oxford: Oxford University Press, 1910.

Bilefsky, Dan. "Inky the Octopus Escapes from a New Zealand Aquarium." *New York Times*, April 13, 2016.

Carls-Diamante, S. "The Octopus and the Unity of Consciousness." *Biological Philosophy* 32 (2017): 1269–87.

Cole, Jonathan. *Pride and a Daily Marathon*. London: Gerald Duckworth & Co., 1991.

———. *Losing Touch: A Man Without His Body*. Oxford: Oxford University Press, 2016.

Finger, Thomas E. "Evolution of Taste from Single Cells to Taste Buds." *Chemosense* 9, no. 33 (2009): 1–6.

Godfrey-Smith, Peter. *Other Minds: The Octopus, the Sea and the Deep Origins of Consciousness*. New York: Farrar, Straus and Giroux, 2016.

———. "Octopus Experience." *Animal Sentience* 26, no. 18 (2019): 270–75.

Graziadei, Pasquale. "Muscle Receptor in Cephalopods." *Proceedings of the Royal Society of London B* 161 (1965): 392–402.

———. "The Nervous System of the Arms." In *The Anatomy of the Nervous System in Octopus Vulgaris*, edited by J. Z. Young. London: Clarendon Press, 1971.

Gutnick, T., et al. "*Octopus vulgaris* Uses Visual Information to Determine the Location of Its Arm." *Current Biology* 21 (2011): 460–62.

Guttfreund, Y., et al. "Organization of Octopus Arm Movements: A Model System for Studying Control of Flexible Arms." *Journal of Neuroscience* 16, no. 22 (1996): 7297–307.

———. "Patterns of Motor Activity in the Isolated Nerve Cord of the Octopus Arm." *Biological Bulletin* 211 (2006): 212–22.

Hochner, Binyamin. "Octopuses." *Current Biology* 18, no. 19 (2008): R897–98.

"Inky's Done a Runner: The Great Octopus Escape." *New Zealand Herald*, April 13, 2016.

Lee, Harper. *To Kill a Mockingbird*. Philadelphia: Lippincott, 1960.

Mather, Jennifer. *Octopus: The Ocean's Intelligent Invertebrate*. Portland, OR: Timber Press, 2010.

Montgomery, Sy. *The Soul of an Octopus: A Surprising Exploration into the Wonder of Consciousness*. New York: Simon & Schuster, 2015.

Nagel, Thomas. "What Is It Like to Be a Bat?" *Philosophical Review* 83, no. 4 (1974): 435–50.

Nesher, N., et al. "Self-Recognition Mechanism Between Skin and Suckers Prevents Octopus Arms from Interfering with Each Other." *Current Biology* 24 (2014): 1271–75.

Roy, Eleanor Ainge. "The Great Escape: Inky the Octopus Legs It to Freedom from an Aquarium." *Guardian*, April 13, 2016.

Sacks, Oliver. "The Disembodied Lady." In *The Man Who Mistook His Wife for a Hat*. London: Picador, 1986.

Sherrington, Charles. *The Integrative Action of the Nervous System*. New York: Scribner, 1906.

Sumbre, G., et al. "Control of Octopus Arm Extension by a Peripheral Motor Program." *Science* 293, no. 5536 (2001): 1845–48.

Wells, M. J. *Octopus: Physiology and Behavior of an Advanced Invertebrate*. New York: Springer, 1978.

Wittgenstein, Ludwig. *On Certainty*. Malden, MA: Blackwell, 1969.

Zullo, L., et al. "Nonsomatotopic Organization of the Higher Motor Centers in Octopus." *Current Biology* 19 (2009): 1632–36.

The Duck-Billed Platypus: An Afterword

Eagleman, David. "The Umwelt." In *This Will Make You Smarter: New Scientific Concepts to Improve Your Thinking*, edited by John Brockman. London: Black Swan, 2013.

Hall, Brian. "The Paradoxical Platypus." *BioScience* 49, no. 3 (1999): 211–18.

Moyal, Ann. *Platypus: The Extraordinary Story of How a Curious Creature Baffled the World*. London: Allen & Unwin, 2001.

Shaw, George, and Frederick Nodder. *The Naturalist's Miscellany, Part 10*. London, 1799. Plates 385–86.

Shaw, George. "The Duck-Billed Platypus." *General Zoology or Systematic Natural History*. London, 1800.

von Uexküll, Jakob. *Innenwelt und Umwelt der Tiere*. Verlag von Julius Springer, 1909.

INDEX